Engineering, economics and fisheries management

Engineering, economics and fisheries management

G C Eddie B Sc, C Eng, F I Mech E, F I Mar E
F R I N A, F Inst R

Fishing News Books Ltd
Farnham · Surrey · England

This is a Buckland Foundation Book,
one of a series providing a permanent
record of annual lectures maintained
by a bequest of the late Frank Buckland

Eddie, G. C.
Engineering, economics and fisheries
management.—(Buckland
Foundation book)
1. Fishery management
2. Fisheries—Economic aspects
I. Title II. Series
338.7'727 SH328

ISBN 0-85238-127-1

Published by
Fishing News Books Ltd
1 Long Garden Walk
Farnham, Surrey, England

Typeset by Paddockglade Ltd, Godalming, Surrey
Printed in Great Britain by Page Bros (Norwich) Ltd, Norwich

Contents

Plates

Figures

Preface

For the subject of the 1982 Buckland Lectures, I chose engineering, economics and fisheries management. In justification of my presumption in straying into fields in which I am no expert, I can plead that Buckland himself seldom hesitated to pronounce upon any and every subject that he felt to be of interest and importance[1]. I have drawn upon my experience of working in various parts of the world as a consultant to FAO and other public-sector agencies, on a wide variety of topics, from investment in the fisheries to the management of research vessels, and from fisheries administration to the harvesting of krill; I have also drawn on my earlier experience as a development engineer at the Torry Research Station and as a director of technical development with the White Fish Authority.

Some of the things to which I draw attention are so much a part of the every-day life of people in the fishing industry that they themselves take them for granted and assume that they can go without saying. I make no apology for including them in a book that deals mainly with policies and activities in the public domain. Less familiar are the economic and practical constraints within which the technical development of the fishing industry has to take place. Nor are the effects of some individual investment decisions upon the problems of fishery management as well understood as they should be.

It is my hope that this book will help to ensure that various peculiar techno-economic characteristics of the fisheries are not overlooked when formulating and administering schemes of management and government support. I cannot pretend to provide any answers to the problems of fisheries management, but I hope to stimulate discussion which may, in turn, lead to new insights. If, to any of those who have to earn their daily bread by grappling with these extremely complicated matters, it appears that I have added to the confusion, I can but offer my apologies. My hope and intention are to try and ensure that certain dimensions of the problems are noticed. To this I add the plea that all who may be capable of contributing to the solution of the problems of fisheries management are given the opportunity to do so.

Early versions of the manuscript were read by Basil Parrish, CBE, Lars-Ole Engvall, Ehrhardt Ruckes and Skipper Willie Campbell, MBE; I am grateful to them for a number of wise comments and helpful suggestions. The opinions expressed in the text are my own, as is the responsibility for any errors.

G C Eddie B Sc, C Eng, F I Mech E,
F I Mar E, F R I N A, F Inst R
Fochabers, Scotland
December 1982

1
Introduction: fisheries technology and fisheries management

The fisheries embrace at least as great a diversity of sciences and technologies as any other single field of human activity. The modern offshore fishing vessel, with its specialized naval architecture and marine engineering; complex fishing gear; elaborate heavy mechanical equipment, electronics and hydraulics; specially-developed refrigerating plant, electronic and acoustic equipment, is so sophisticated that it costs more to build per ton of displacement than any other common kind of ship, with one exception – the ship of war. Yet it is only in the last thirty-odd years that professional engineers have been employed in significant numbers in fisheries research and development. On only one previous occasion has the Buckland Professor been a professional engineer.

The advent of the engineers coincided with, and greatly reinforced, the great expansion of world fisheries that began in the 1950s. Their contribution can be measured by comparing the

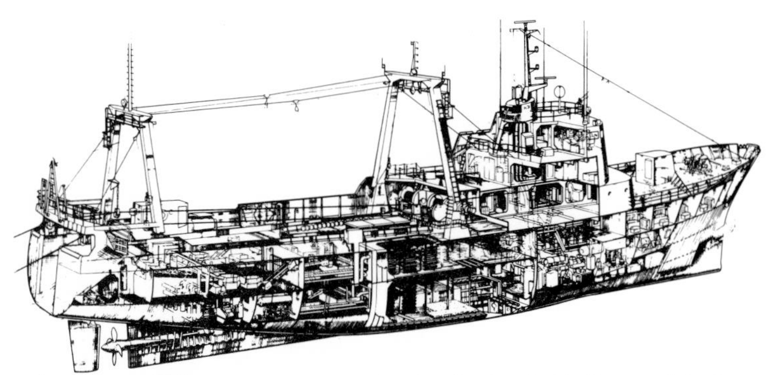

Plate 1 This cutaway perspective drawing by Robin Perry of the freezer trawler Saint Jasper illustrates the complexity of modern deep-sea fishing vessels *(Courtesy Robin Perry Technical Illustration)*

specifications and performances of modern fishing vessels with those of the corresponding craft of thirty years ago, and by a study of the statistics of the size and composition of the world's fishing fleets[2]. Between 1950 and 1970, world fish production trebled[3]. During that period, and up to the present time, the engineers have been preoccupied for the most part with more-or-less straightforward technical development and with research in direct support of such work, although it should be borne in mind that this has included operations analysis aimed at improving system performance or efficiency, the analysis of investment proposals, the evaluation of costs and benefits and so on[4].

Engineering is interdisciplinary and always involves people. Ships, like other engineering products, perform most effectively and economically if those responsible for their design have been able to familiarize themselves thoroughly with the application, that is to say, with how the ships and equipment are used and the environment in which they have to work. The engineer ought therefore to maintain close contact with representative users. In the course of developing new or improved methods and equipment for the harvesting, preservation, distribution and marketing of fish, engineers have perforce to come into very close working relationships with fishermen, managers of fishing fleets and processing plants, and all the other kinds of people who work in the fisheries industry and in its supporting industries and services. Some engineers have had to become familiar with conditions at sea on board a variety of commercial fishing vessels in various areas of the ocean; others with the markets, processing plants, cold stores, depots, retail outlets and other parts of the distribution networks on shore. They have thus become very much aware that the fisheries industry, and the people in it, possess certain peculiar characteristics and qualities that distinguish the fisheries from other industries. Some of these peculiarities arise out of the environment in which the industry has to work; some because the harvesting of fish is essentially hunting, and requires very special skills and technology; some arise out of the perishable nature of the product.

In the field of fisheries management, the perennial problems of conservation of resources, allocation of catches between competing groups, the profitability of the fishing fleets, and the effective utilization of the landings, are difficult enough even when there are only gradual changes in methods and equipment and in the market situation, and the size and composition of the fishing fleets are stable. In periods of rapid expansion, or innovation, the problems are exacerbated. In the literature on fisheries management and economics, technology is discussed mainly in terms of its effects upon costs of fishing and upon the intensity and mobility of the fishing effort. Much of the activities of the engineers have indeed had the result of increasing the fishing effort that can be exerted by a vessel of a given size, type and power, or by a given number of men. Indeed, a continual process of technical development has made it difficult to measure or to define effort, except in terms of its effects. Other aspects of technology, however, and allied matters – such as the importance of individual skills – are less often noticed. At first sight, this is surprising: after all, the fishing industry is characterized mainly by its ships and other equipment and by the fisherman's special skills. Those directly concerned with equipment and skills are, however, only recent arrivals in the field of fisheries research and development; most of the theoretical debate on fisheries management has hitherto been conducted by scientists and economists. The apparently never-ending discussions on allocation of catches, over-fishing, profitability and the need for better marketing, while they take note of some of the peculiar techno-economic characteristics of the fisheries, seem often to ignore others. The wide range of performance of individual vessels in the fleet is seldom mentioned. The fairly voluminous literature on fisheries management also pays scant attention to such matters.

The remainder of this book is devoted partly to listing some of these peculiarities and discussing their significance for fisheries management, for the organization of research and development and for the role of engineers and other technologists in government. Finally, it offers some views on the nature of the technical effort required if the full potential of the world's fisheries resources is to be realized.

2
Techno-economic considerations

Historical

The design, manufacture and use of fishing gear is one of mankind's oldest technologies: fishing spears, hooks, gorges and sinkers are found in late upper Palaeolithic and post-glacial Mesolithic occupation sites[5]. The fisheries are the only remaining instance of food production by hunting or gathering that is practised on a world-wide scale and is economically significant in world terms. Despite the long history, the full potential has not yet been realized (it is believed to be from about one-and-a-half to over three times present yields[6]) no doubt because of the daunting technical problems of harvesting, at economic cost, the less accessible resources in the oceans, problems which are only now becoming capable of solution.

Nevertheless, as long as eight thousand years ago, people living in the north of the Iberian Peninsula, who had for long enjoyed a diet that included littoral molluscs, salmonids and other species that could be gathered or captured near the water's edge, were already beginning to take offshore bottom-dwelling marine species from the Bay of Biscay[7]. Whether this was a response to declining yields from the littoral and estuarine resources, or to growth of the human population, or was a spontaneous development of existing technology, is not clear. Much the same was happening in Peru.

What is certain is that venturing further afield, and into deeper and more exposed waters, is not a recent phenomenon. Fishing therefore probably became a specialized activity at an early period. Techniques of preservation may also have been well developed at an early stage, because although, no doubt, fish could be caught very easily in prehistoric times even at the water's edge [as in Newfoundland in recent times and the Falkland Islands (Mal-

Plate 2(a) A traditional fishing craft, the kattumaram, in Sri Lanka (*Photo: FAO*)

Plate 2 (b) Built, like the kattumaram, for the hunting of fish, vessels like this German stern trawler were developed in the 1950s (*Photo: FAO*)

vinas) even at present] such abundance and availability are often seasonal. Fish, however, are easier to preserve by drying, salting or smoking, and to transport, than any large species of terrestrial animal that might have yielded an equal weight of edible flesh for the same effort in its capture.

The Phoenicians and Carthaginians were active fishermen[8] and by about 500BC fillets of sea bream were, it seems[9], transported from the Western Mediterranean to Greece. The taste of Imperial Romans for British oysters is well known. The trade in salted and dried fish from northern and western Europe to the Mediterranean lands, so important throughout most of recorded history, had already begun in classical times, when the Continental Celts exchanged these and other commodites for wine[10]. By about a thousand years ago, in the so-called Dark Ages, the Northmen had developed really seaworthy, ocean-going ships; it is tempting to speculate that this development, with all its consequences, was based in part upon the existence, off the shores of northwest Europe, of abundant stocks of fish: a convenient supply of food, including incidentally rich sources of vitamin D, for people living in these high latitudes.

Historical records indicate that the problems of fishery management were already evident in the European Middle Ages, leading rulers and governments to take action of various familiar

Plate 3 The caballito del mar was in use in Peru centuries before the advent of the Europeans. Compare *Plate 5* (*Photo: FAO*)

kinds. The Icelanders were complaining about the behaviour of fishermen from England some five hundred years ago and more[11], in terms very similar to those used during the recent cod wars. In the year 1592[12] the Council of the Burgh of Aberdeen severely disciplined two fishermen who had taken violent and destructive objection to a new type of fishing gear developed by one of their colleagues, which the Council considered to be 'a useful invention and most profitable for the common good'.

In August 1588, the Invincible Armada and its English opponents were not the only fleets proceeding up the Channel: some of the deep-sea fishermen of Dieppe and Le Havre sailed through the hostile fleets as they made their way home from their annual campaign on the Grand Bank and Newfoundland Banks[13]. Distant-water fishing is much older than mechanical propulsion. All that was required was a comparatively large and seaworthy ship, the ability to navigate over long distances, an effective means of preservation (salting or drying), a willingness on the part of the men of the fishing communities to endure long

Plate 4 The task of mending nets has been a familiar one for thousands of years (*Photo: FAO*)

absences from home, and acceptance of this by their womenfolk.

In its basic skills, methods, equipment and processes, the fisheries industry has changed remarkably little. Even in the most advanced fishing industries the basic methods and equipment are recognizably the same as those of the less advanced, and not all that different from those of the Upper Palaeolithic, except in three respects: the use of radio navigation aids; the use of underwater acoustics, and the application of mechanical power, including mechanical refrigeration[14].

Mechanization and its consequences

The use of mechanical propulsion (or, as the international development agencies prefer, 'motorization' or – less precisely – 'mechanization') confers superior mobility: the vessel is much less at the mercy of wind and current; often it also confers higher speed. Some of the consequences are best described by looking at specific examples.

The diverse and very important fisheries off the southwest coast of India are still prosecuted mainly by craft of traditional types, propelled by sail, oar and paddle. The fisheries are, therefore, perforce, opportunistic and seasonal: the fishermen have to wait for the various stocks of fish to come within their very limited radius of operation, during the annual cycles of migrations. Various combinations of craft and gear are employed according to species and season. At some seasons, no fishing can take place, because the craft are not capable of following the shoals, or of making an active search for them. For the same reasons, fishing opportunities vary from district to district and some communities are much more successful and prosperous than others[15]. In very recent years, some mechanized trawlers and gillnetters have been introduced; these will be mentioned again a little later.

On the Batinah coast of Oman, where the level of technology is otherwise similar to that in southwest India, the use of the internal combustion engine to propel craft of otherwise traditional design is now widespread. The smaller and less seaworthy types of craft are still confined to opportunistic fishing close to shore and off the home village. The bigger, mechanized huris and sambouks, however, can at times be seen concentrated in large numbers inside one or two square miles of sea, where presumably the fish are temporarily most abundant, and these craft will have come from several villages scattered along many miles of coastline. Such concentrations would have been unlikely in the days before the introduction of mechanical propulsion and would have been possible only if wind and current were favourable and if the fishermen were in the habit of migrating along with the shoals, sometimes far from home. Thus, the added mobility conferred by mechanical propulsion may have allowed larger numbers of vessels to congregate where the best fishing is, and hence an increase in effective fishing effort much greater than might at first be supposed.

Plate 5 Mechanized fishing vessels in Callao, Peru, waiting for a run of anchoveta. Compare *Plate 3* (*Photo: FAO*)

A capability for hunting and intercepting migratory species, at greater distances from the coast than were practicable in the days before mechanical propulsion, improves the chances of success for the bold and enterprising individual fisherman, at the cost of consuming fuel and other resources, and perhaps also at the expense of his colleagues. An extreme example is the marine fishery for salmon off northwest America: in theory, at least, it should be possible to replace most of the seiners, trollers, gillnetters, barges and transport vessels by a few sluice-gates or traps on the rivers, a strategy still employed by the bears and formerly by a flourishing and prosperous Indian society. The additional costs to the community of the present system are obvious; the benefits include the preservation of skills and other human qualities that might otherwise be lost.

In Bangladesh, small mechanized gillnetters have recently been introduced which operate in the upper part of the Bay of Bengal, in shallow water a few miles off the coast[16]. Part of their catches is of anadromous species that would previously have been taken by fishermen in the river delta, nearer to the centres of consumption; part of the rest might have been taken in the calm-weather season by the fixed behundi-nets just off the mouths of the rivers[17]. The real benefit of introducing the mechanized vessels may be not so much any immediate improvement in volume and regularity of supplies, but the development of sea-going skills on which an expansion into offshore waters can be based.

Manpower and productivity

What may be described as the first phase of mechanization – the adoption of mechanical propulsion – is, as we have just seen, still in progress in the developing countries. For the fishing nations of the northern temperate latitudes this phase began over a hundred years ago and was virtually completed during the first half of the twentieth century.

More recently, in these countries, mechanization has been carried a stage further, in order to reduce the physical effort required during the handling of the fishing gear and the catch, and to improve manpower productivity[14]. It was only by increases in productivity that the British distant water fleet was able to meet the national objective of maintaining landings at a roughly constant level throughout the 1960s and into the 1970s, in spite of a continual fall in the number of men able and willing to go to sea in the big trawlers. One result of reducing both the physical effort and the mental strain of distant-water fishing was that the older and more experienced fishermen found it possible to continue in active service to a greater age than had previously been normal[18].

The growing scarcity of fishermen has affected all the more affluent of the major fishing nations, and there is a constant pressure to improve productivity. In so far as this has meant investing in more, and more sophisticated, equipment, it has been achieved at the cost of increased energy consumption.

Socio-economic factors

Mechanization is not always a simple matter of mounting programmes of demonstration and training around some pieces of equipment already in existence and commercially available. Many years of effort, for example, have gone into attempts to develop mechanized fishing boats capable of operating off beaches beaten by heavy surf, but in one particular region, the east coast of India, signs are only now apparent that a satisfactory solution may be on the way[19]. The objective of this project is to develop a fishing craft with greater speed, range and carrying capacity than the traditional log-raft, the kattumaram, so as to enable the fishermen to exploit the resources that are believed to exist further offshore than the traditional craft can normally operate. In order to cover the additional costs arising from mechanization, and also to improve the fishermen's net earnings, the new craft must be capable of carrying more fishing gear and a bigger catch. The immediate consequence of installing an engine on a kattumaram, however, is to reduce the payload, so a craft with improved carrying capacity has to be developed, still light enough to be manhandled up the beach, and sufficiently strong and otherwise capable of withstanding impacts and capsizes when coming in to shore. This represents a considerable, and to some engineers, a fascinating technical problem, especially when the solution has to be pro-

Plate 6 (a) Kattumaram on beach in Kerala. Installing an engine reduces the payload (*Photo: FAO*)

Plate 6(b) Kattumaram coming ashore through the surf near Madras (*Photo: BOBP*)

duced within the practical constraints of locally-available materials and low finished cost. For a while it seemed to at least one onlooker that a craft better than the traditional kattumaram might only be possible if resort were had to aero-space materials and methods of construction. If that had been so, a better solution would have been a vessel built of local materials, but much larger, so as to achieve a more favourable deadweight ratio; such a vessel, however, would have required the provision of mechanical aids to handle it on the beach, or the building of a breakwater, and would have meant developing new patterns of ownership and cooperation in the communities of the fishing villages. In the event, some promising designs have been developed that avoid these latter, socio-economic problems, being straightforward replacements for the kattumaram. These are currently undergoing extended trial in practical fishing conditions[20].

Previous attempts to introduce mechanized craft were sometimes technically successful but did not meet with acceptance for rather different reasons. Pere Gillet[21] recalls '. . . one 24ft surf-crossing beach boat (FAO design) equipped with outboard motor . . . demonstrated brilliantly her surf-crossing capabilities at Muttom Beach (South-West Tamil Nadu) . . . Manned by a local crew under a Canadian master fisherman, the boat went fishing up to the Wadge Bank and recorded bumper catches, which provoked the local population, who accused the Project of emptying the fish wealth of the sea'.

Some African communities, according to Carleton[22], still believe that it is wrong to take from nature more than one's immediate requirements; such communities, presumably, have no experience of winter, or of the problems of preserving a mammoth or a shoal of herring when these represent more food than the local community is capable of consuming in the immediate future. Pere Gillet[21] also observes: 'Mechanization of fishing operations implies a basic change in the pace of activity. Fishermen operating non-mechanized craft work less during the "peak season", cover their needs and enjoy life: it is the time for social functions, the marriage season, entertainment and pilgrimage. But during the lean season these same fishermen work very hard to earn a few rupees needed for their sustenance. Mechanization requires a different attitude: one has to work hard during the "peak season" to have good returns for the capital invested, but all activities may have to be stopped during the "lean season" due to the inability to cover costs of fuel et cetera'.

Plate 6(c) Prototype replacement for the kattumaram, designed by the FAO/SIDA Bay of Bengal programme (BOBP), undergoing trials near Madras (*Photo: BOBP*)

Reliability

However, fuel costs are giving less cause for concern to some fishermen than the costs of repairs and maintenance, especially labour costs, and the virtual losses arising from the loss of fishing time occasioned by breakdown. There is a need for mechanical equipment that is more reliable, and that is capable of being rapidly dismantled, repaired and re-assembled, using simple, locally-available skills and facilities. This need is not confined to the sophisticated vessels of the advanced fishing nations; in the developing countries there is the same need for simple, reliable,

Plate 7(a) Ouidah, Benin People's Republic. Outboard engines are used to propel dugout canoes. They can become a financial burden unless . . . .(*Photo: FAO*)

easily-repaired equipment, and perhaps especially for an outboard engine more suited to continuous use in commercial fishing conditions than the models mass-produced for the sport and pleasure market. There are two problems: the comparatively small size of the fishery industry as a market for specialized engineering products; and lack of a statement of operational requirements in technical terms comprehensible to a design engineer. The same was true of diesel engines and hydraulic equipment in the developed countries a generation ago, as noted in Chapter 4. More recently, the author has come across a situation where engines had been introduced for propulsion of traditional fishing craft, without adequate facilities for maintenance, provision of spares, or training; there was about an even chance of earning enough extra money to cover the purchase price of the engine before it broke down through damage to main bearings or crankshaft.

Plate 7(b) .... adequate training and facilities are provided for maintenance. An FAO-trained mechanic services an engine at Lome, Togo (*Photo: FAO*)

Mobility of fleets

The greater mobility conferred by mechanical propulsion had the result that more of the catch was landed in the 'fresh' form, and in a fresher condition, than in the days of sail and oar – accomplished sometimes by systems of fast carrier vessels plying between the fishing grounds and the shore markets, as formerly operated in the North Sea, and still do in the Pacific salmon fishery and off the mouths of the Ganges and Brahmaputra[17]. The crews also saw home more frequently.

Mobility is also an insurance against failure of the local fishery or of the traditional market: one of the situations which give rise to appeals for help addressed to rulers and governments, is when a fishing community is faced with a reduction in the abundance or availability of a stock of fish upon which they have hitherto depended, or with the failure of the market for their traditional product, and when moreover they do not possess the knowledge and skills, or the financial capital, that would enable them to re-equip and to re-deploy their effort. The value of flexibility and versatility then become apparent, but they have to be paid for in higher capital investment; in vessels with greater range, endurance, speed and seaworthiness; in having available more different kinds of fishing gear. All of this implies higher costs. As an example, the optimum boat for fishing inside the Bay of Fundy will probably be of more modest size and power than the vessel that would be preferred for fishing outside the Bay in the open Atlantic; the fishery in the Bay has been known to fail; investment in the larger boat would be an insurance against this eventuality but incurs additional costs. It is not realistic to argue that the fisheries in the Bay itself would never fail if only they were properly managed; there is too high a probability of failure for reasons, other than over-fishing, that are outside man's control or ability to predict. The choice is therefore between providing temporary assistance from time to time, when fishing is poor, financed by a levy or from public funds, or acceptance of the additional costs of a fleet composed of vessels sufficiently versatile to be re-deployed when necessary, always supposing that there is another fishery on which they can be re-deployed.

The optimum vessel

The foregoing suggests that it is not such a straightforward matter as may at first appear, to discern what is the optimum vessel for a given fishery. Fishing vessels can have useful service lives of fifteen or twenty years or more, and it is all too easy to be overtaken by events. In the 1950s and 1960s much work was done in the United Kingdom on the optimization of the design of the new freezer trawlers then being built to fish in the distant waters of the North Atlantic and Arctic. The vessels they were to replace had stowed their catches in crushed ice, and the earliest-caught fish could sometimes be in an advanced state of spoilage by the time the vessel landed its catch; high speed on the long passage home was then an advantage. High speed, however, required very high propulsive power, with concomitant high capital and running costs, and only a fraction of the power was required when actually fishing, with the designs of trawl then available. It was argued that for the new freezer trawlers, with the catch quick-frozen and stowed at low temperature, there was no longer an advantage in high free-running speeds[23]. Some vessel owners accepted this argument, and built freezer trawlers of what was regarded as optimal design as regards hold capacity and speed, as indicated by sophisticated computer simulations[24]. Other owners built freezer trawlers that could attain speeds equal to, or higher than, those of the wet-fish trawlers they replaced, with propulsive powers far in excess of those required when fishing (with the gear then available); clearly these vessels were not the most economical in the conditions then obtaining.

Some time afterwards, Schaerfe, at the Institut fur Fangtechnik in Hamburg, perfected the single-boat aimed midwater trawl[25]. Such trawls could be designed to absorb all the propulsive power that could be developed by any trawler in service: the more power available, the bigger the trawl that could be towed (or the finer the mesh). Those vessels built for high speed, having greater propulsive power, could take most advantage of the new gear, whenever conditions indicated its use was worthwhile. Some years later, however, the introduction of 200-mile limits sharply reduced the scope for all European freezer trawlers in the North Atlantic.

This history illustrates the possible mismatch, hinted at earlier, between the likely physical life of a fishing vessel and the 'life' of the fishery: ships have to be built to withstand the sea, and therefore perforce can remain seaworthy for many years; if they could be built like Japanese paper houses, then the problems arising out of particular fish stocks no longer being accessible, or no longer being in sufficient abundance, might not loom so large. Since we have even less influence over the marine environment than we have over the abundance and availability of the fish stocks, there is a premium on versatility in the vessels and on the ability to re-deploy them, which requires not only suitable equipment but also a wide range of knowledge and skills, as well as access to alternative stocks.

Size, catching power and costs

However, the more mobile and versatile a fishing vessel, the greater will be its capital and running costs, other things being equal. In order to cover the greater costs, it must catch more fish (of given unit value). It is widely assumed by laymen, including the public press and radio, legislators, and even some administrators, that the bigger the vessel, the more fish it is likely to catch in a given time. This is far too simple a view, and it is only partly true. Leaving aside for the moment two very important factors, namely, first, the skill of the skipper and crew, which is discussed later below, and second, the right choice of method of capture (of which a little more in a moment) there are two other important parameters: weather tolerance and propulsive horsepower[26].

As regards weather tolerance, even in a notorious bad-weather area like the North Sea, the 20 to 30m LOA Scottish bottom seiners will, if conditions justify it, continue to fish in just as bad weather as trawlers of 35 to 50m and even bigger. As regards propulsive power, catching potential does in some cases partly depend upon it, as for instance when using the midwater trawl, as mentioned a little earlier, but the relationship is by no means straightforward: it depends upon fish behaviour and distribution, and the fishing situations where there is no alternative to very high powers, and where these are of decisive advantage, are not all that common: the best-known is perhaps the krill fishery of the Southern Ocean. The Scottish bottom seiners, for example, pride themselves on being able to outfish trawlers of three or four times their power. The alleged devastation caused by Russian distant-water fleets is not due to the size and power of the vessels (individual rates of catch are often not remarkable) but to the large number of ships deployed on the fishing grounds.

What sheer size confers is range and endurance: the ability to reach fishing grounds where the stocks are more abundant and where catch rates are higher than they are nearer home. There will be a continuing requirement for large ocean-going fishing vessels despite the general adoption of 200-mile limits, not only to take the truly oceanic species – the tunas and oceanic squids – but also to harvest the abundant resources lying off coasts that are too inhospitable, or too remote, to justify the founding of local settlements of fishermen: northern Labrador, Patagonia, the Antarctic.

In any particular fleet, the vessel that produces the biggest annual landing or achieves the highest gross earnings is not always the most profitable: this is often a vessel of a size and power nearer to the minimum that is operationally acceptable – nearer to the smallest and simplest vessel that will do the job. This is in conflict with what was said earlier about the need for versatility and mobility, but the larger and more expensive to run that a vessel is, the more the skipper will be compelled to seek especially good fishing opportunities in order to attain the high rates of catch needed to make such a vessel profitable.

Plate 8(a) A pirogue on a beach near Dakar in 1980. Such craft are of limited range but incur modest costs (*Photo: FAO*)

There is a tendency to specify new vessels to match the abilities of the better skippers and crews, and sometimes of such size and power that only the very best skippers and crews possess enough knowledge, skills and energy to make them pay. The larger and more powerful a fishing vessel is, the higher the rates of catch that have to be attained to cover its costs. This is part of the reason why certain ocean-going fleets have indulged in 'pulse' fishing: finding a stock of fish that is underfished, and fishing it very hard for as long as catch rates may justify. It is not at all clear whether there is enough known of the biology, life cycles and behaviour of any fish species, to allow pulse fishing to be managed in a conservative way. Another stock or stocks would have to be accessible, on to which the fleet can be re-deployed while the first stock is recovering, if it will. This calls for large and mobile vessels which need fishing opportunities offering high rates of catch, which are more likely when stocks have been underfished for some years . . . and so it would go on.

If, on the other hand, it is desired to manage the fishery on a

Plate 8(b) Deep-sea fishing vessels in Dakar in 1980. Costs are much higher than for the pirogue; catch rates have to be correspondingly greater (*Photo: FAO*)

Plate 8(c) Russian factory trawlers off Dakar in 1980. Such vessels can range world-wide; the problem is to achieve high enough rates of catch to cover the very high costs (*Photo: FAO*)

quasi-steady-state basis, as is assumed in so much of the literature and in many existing regimes of management, then more attention should perhaps be paid to ensuring that there is not over-investment in the sense of individual vessels being too big, too powerful and too costly. Experience shows that this cannot be left entirely to market forces. One reason, already mentioned, is that it is desirable to take the precaution of having a vessel that is capable of being re-deployed in the event of failure of the fishery or of the market for the product; choice of size, power and equipment then becomes very much a matter of judgment, and the man who plays safe penalizes himself financially, unless the worst actually does happen. Another reason is that any attempt to discern what is the optimum vessel for a given fishery, in terms of size, power and equipment (optimum as distinct from the merely adequate, technically and operationally speaking) is made difficult by the very wide range of performances of individual skippers and crews, a topic to which we will return later.

The larger and more mobile vessels in a national fleet may fish in waters far from the home port, and sometimes they may fish for species that are not highly prized in the communities where the fishermen have their homes and families. In such circumstances, the systems of fishery management by voluntary limitation of effort, exercised by local fishing communities, which Troadec[27] believes on historical evidence to be the most practical and effective systems of fishery management yet devised, may not be completely effective in modern circumstances. The adoption of 200-mile limits is not in itself enough in coastal states where the Continental Shelf is big: a vessel capable of operating 100 or 200 miles or more up and down the coast, is a highly mobile vessel.

Energy costs

The application of mechanical power made certain methods of capture much more effective than they had been in the days of sail, oar and paddle: trawling, purse-seining, even deep-sea long-lining. Much has already been written about these developments[14]. Attention is now focussed upon the concomitant consumption of energy derived from fossil fuels. Nowadays, for mechanized fishing vessels, the cost of fuel is often the biggest single cost: bigger than labour costs or costs related to capital. This has tended to obscure the comparatively high productivity that it is possible to attain in the fisheries: when there is a reasonably good fishing opportunity with the fishing grounds reasonably nearby, and a correct choice is made of method of capture, fishing can sometimes produce more, sometimes much more, edible protein, per unit of fossil fuel consumed, than can most systems of intensive mechanized agriculture depending upon chemical fertilizers[6]. A rise in fuel prices, however, affects the fishing fleets immediately, whereas it takes a year or more to work its way through the farming system.

The increases in the prices of fossil fuels in recent years have led to re-appraisals of the extent to which mechanization is desirable in the fishing fleets of some developing nations, and how it should be applied. In the fields of mechanical prime movers and propulsion systems, there is no obvious alternative option that the new situation makes especially attractive to the fishing industry. Minor adjustments and modifications to the design of hulls and machinery will produce some savings, but the biggest opportunities for improvements in fuel economy and productivity in the fisheries are likely to come from changes in choice of method of capture, improvements in the level of skill of fishermen, and changes in operational practices to achieve more effective utilization of the vessels.

Other costs

In at least one case where the profitability of a mechanized fleet leaves a good deal to be desired, high costs of fuel have been blamed when, in fact, the cause seems to be rather different. The mechanized trawlers and gillnetters operating off the coast of Kerala, in southwest India, are no longer operating, on average, at a profit[15], while much of the traditional, non-mechanized fleet still does, on average, make a profit. The obvious conclusion is that the difference is due to mechanization and, more specifically, to the increased costs of fuel in recent years. However, the average earnings of a deckhand on the mechanized vessels are very much higher than those of the average traditional fisherman. The manning scales and systems of sharing the earnings between

Plate 9 Mechanized fishing boats undergoing maintenance in Kerala. Compare *Plates 6* and *14b*; mechanized craft incur higher costs per crew member (*Photo: FAO*)

vessels and crew in the mechanized fleet were fixed when the mechanized fishery was virtually a virgin one, catch rates were high and fuel costs were lower than they are now, and pay too much regard to traditional practices and not enough to the fact that the mechanized vessels represent a higher investment per crew member, and incur costs which a traditional vessel does not. One result is that the vessels are often laid up for annual overhaul during the season when the fishing is best, because only then is the cash flow to the vessel, after paying the crew, sufficient to cover the costs of maintenance.

The more capital-intensive the operation, and the more technically-sophisticated the vessel, the higher the proportion of gross earnings that will have to go to the boat. At the same time, if the investment is sound, the incomes of individual crew members should be as good as they were before, or somewhat better. To achieve this, however, it may be desirable to reduce the manning scale, and this may only be achievable, operationally speaking, by installing more equipment and incurring yet further capital and running costs.

In the contexts of fisheries management and national fisheries policies, little attention has yet been paid to this particular consequence of adopting ever-more complex technology and building ever-costlier vessels: systems of sharing gross earnings, and manning scales, are seldom reviewed, questioned or revised.

Few, if any, would begrudge the fisherman what he earns, in conditions that are always harsh and often dangerous. Situations broadly similar to that just described are familiar in the fishing industry: another example that comes to mind is that when a trawler in the British distant water fleet was just covering her costs in the mid-1950s, the skipper's share of the annual gross earnings was already equal to the salary of the Prime Minister at that time[28].

The market

What these examples illustrate is that in many of the world's fisheries, prices are not under the control of the vessel operators. They are subject to several sorts of constraint, including social attitudes to fish as food, or to certain species, the availability of competing foodstuffs, and various imperfections of the market and the marketing system. A discussion of most of these factors is beyond the scope of this book, but there is something that must be said, arising out of two particular attributes of fish that are particular concerns of the technologist: its perishability, and the unpredictability of supplies.

Fish is highly perishable; most of the catch has to be dealt with in one way or another within a few hours of landing. Even when frozen at sea, or salted on board the catching vessel, it must undergo further processing on shore: Burney's dream of producing consumer packs at sea has been realized, so far, only to the extent that some canned fish products are produced on board Russian factory ships. Supplies at the ports, however, fluctuate from day to day and are unpredictable for more than a day or two ahead. There is therefore strong interaction between the two sets of operations: catching and landing the fish, and processing and distributing it on shore; this interaction is illustrated for example by the way in which prices tend to vary in inverse ratio to daily landings in those ports where the catches are sold by auction; prices are insensitive to quality except in periods of glut.

The catching and landing of fish is therefore just one part of a closely-integrated system for bringing fish in the sea to the consumer's plate. Landing fish on a beach or quay or jetty is not in itself a useful economic activity. It may, therefore, be very misleading to think in terms of, and legislate for, an entity called the fish*ing* industry, and better to think in terms of the fish industry as a single system or set of systems. There can arise situations where the best opportunity for improvement in the wellbeing and prosperity of the fishing fleet may lie in some development in the system of marketing and distribution. The reverse is also true: it is not sufficient to assume that the solution to the industry's problems will lie always or merely in better marketing, or in better physical systems of preservation and distribution on shore.

There is a natural temptation to put too much faith in development on the shore side, because it is more obvious to the layman what kinds of talents, methods and resources offer the promise of improvement, and it is likely that these talents will be commoner and more readily available than those needed to carry out development work at sea. Moreover these talents will be less dangerous, less uncomfortable and less expensive to exercise, than those that are required if we are to ensure improvements in the regularity, quality or volume of the landings, or in the costs of

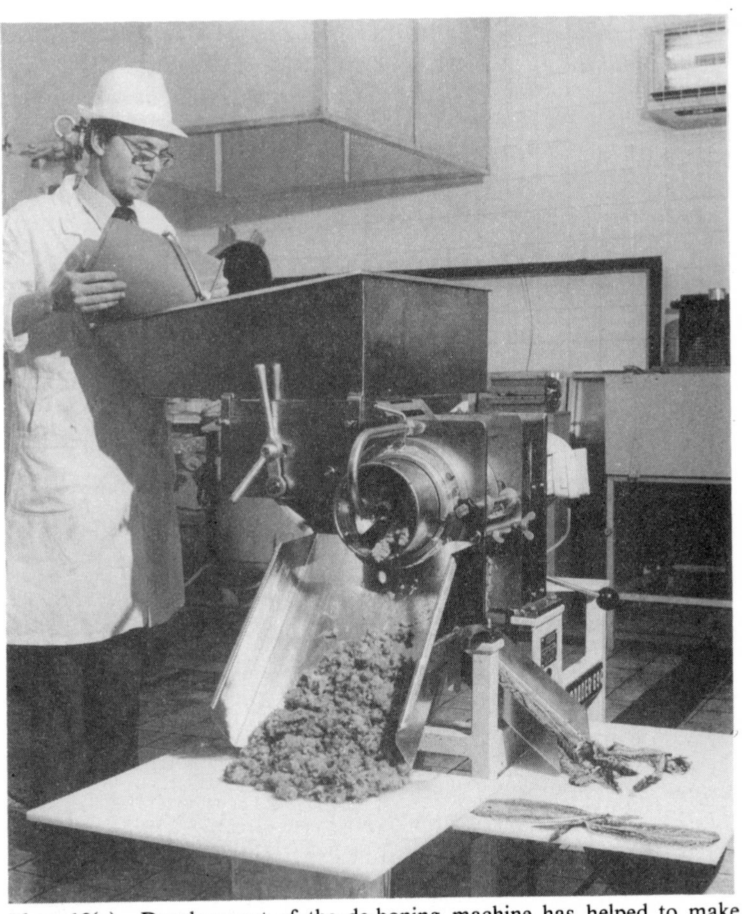

Plate 10(a) Development of the de-boning machine has helped to make possible . . . . (*Photo: Torry Research Station*)

production, or in the earnings and working conditions of the fishermen. This is why, although a prominent feature of some fisheries administrations is an inspection service in the processing plants on shore, sometimes very elaborate, there is much less often an effective organization charged with the task of developing adequate equipment and practices for handling and processing the catch on board the fishing vessel at sea, despite the fact that the spoilage of fish poorly handled and stowed is irreversible.

The fluctuations in quantity, size and species of fish available might be reduced in amplitude by appropriate management of the fisheries; they can also be smoothed by freezing in bulk, salting, drying or canning, in bulk, as may be appropriate. Bulk-processing techniques are still not fully developed. The vagaries of supplies from day to day and season to season might also be of less importance if more advantage were taken of the development, in recent years, of standardized products that can be made from a variety of different fish species and sizes of fish, of which the fish finger, and surimi, are examples[29]. Any or all of these steps would change the precise nature and the magnitude of the marketing problem.

The problem takes a different form in Oman. The fish resources within that country's 200-mile limits are probably the biggest potential source of wealth after oil, but existing craft can fish only within a few miles of the coast, and most do not venture out of sight of land. Because of the demand for fish in the Arabian Peninsula and Gulf, the incomes of individual fishermen, even in the remote south, have in recent years been so high that there is no

Plate 10(b) .... a number of standardized food products that can be made from whatever species are in greatest abundance (*Photo: Sea Fish Industry Authority*)

incentive to invest in a bigger or more powerful boat, and so the major part of the resources is not exploited by the local fishermen and as yet contributes little to the national income.

The role of government

In the world as a whole, very few of the various government departments and agencies responsible for the well-being of the fishing fleets, or for the supplies of fish to the populace, are technically capable of covering the entire system of production and distribution in a comprehensive way. The growth of fisheries administrations and their scientific and technical arms has been *ad hoc* and piecemeal. Rulers and governments have almost always been involved in the fisheries, either because they asserted rights of property, or because they found themselves called upon to settle disputes; during the nineteenth century biological scientists and oceanographers were recruited to study the commercially-important resources and their conservation; early in the twentieth century, the need to ensure supplies of nutritious foodstuffs to urban industrial populations, especially in time of war, brought in food scientists and food technologists; the peculiar circumstances and problems of the fisheries attracted the attention of some economists; by about thirty years ago it became evident that there was a need for the services of professional engineers and allied disciplines to undertake technical development of vessels, equipment, processing plant and systems; more recently still, lawyers have been much in evidence. In the public sector, the activities of these various groups have in many countries been carried out, more often than not, each under the auspices of a separate, more or less autonomous or independent agency, or under separate departments of state.

In the field of marketing, the private sector provided most of the effort in former times, as for example in the early days of mechanized fishing, manufacture of ice and transport of cheap fresh fish to inland centres of consumption by rail; again in the second quarter of the twentieth century, when the practice grew of filleting the fish at the port before despatching it by rail; yet again in the 1950s when the frozen consumer pack and the fish-finger made their appearance. Nowadays, marketing experts are also to be found in the public sctor, at both national and international levels[30].

Rothschild argues[31] that the correct role for the public sector is to carry out research and development aimed at reducing the real costs of production, reducing waste and broadening the supply base: he suggests that the taxpayer should not normally be expected to spend money on those kinds of marketing activities which are aimed primarily at increasing demand and hence price [32]. (The problem remains, however, of organizing effective generic advertising for an industry composed of a great many small enterprises with somewhat divergent interests and views that, owing to various pressures, are often necessarily short-term.)

The rather haphazard organizations and institutional frameworks which have resulted from the historical processes just mentioned mean that, even now, the potential contributions of each of the specialist groups to the central problems of fishery management are only partly recognized and are far from being fully realized. Each of the groups can play a part in alleviating the problems of conservation of resources, allocation of catches, utilization and profitability; a properly coordinated and combined effort might succeed where separate and independent initiatives seem to be inadequate. Some simple examples illustrating this thesis are presented later below, in the course of a discussion of the peculiarities, the potential and the limitations of technical development in the fisheries industry, and its relationship to scientific research. First, however, it is necessary to round out the description of the industry by looking at the importance of individual skills and performance, and other personal qualities.

3
The nature of fishing enterprises

Effective management of fishing fleets and use of fishing craft and equipment call for a variety of uncommon abilities and acquired skills; fishing engenders certain attitudes of mind and modes of behaviour. The range of performance of individual skippers and crews is wide and the implications may be more significant than is usually allowed in the literature on fishery management and economics.

Business management and the fishing industry

Although fishing is the hunting of self-reproducing stocks of wild animals over which man has little control, in a very harsh environment over which he has almost no control, attempts are nevertheless sometimes made to apply principles and systems of business management appropriate to manufacturing industry, including the setting of production targets, and systems of production incentives, for particular fleets, and even individual vessels, that would be more appropriate to the production of radio sets or nuts and bolts. Such cases seem to betray a lack of appreciation that a production target in a fishery can be no more than a declaration of intent, perhaps little more than an expression of hope. This is so even in those cases where the fishery has been surveyed and estimates made of the long-term yield, or forecasts made of the abundance during the forthcoming season.

Nevertheless production incentives must be employed, otherwise performance will be very poor indeed, as was observed by the author a few years ago in one state fisheries corporation operating deep-sea trawlers, the officers and crews of which were paid straight salaries, irrespective of how much or how little fish they caught; since they were based on a foreign port and got regular (if infrequent) home leave, there was not even the incentive of filling the hold quickly in order to get home sooner. In another case, where mechanized trawlers have been fairly recently introduced, the costs of fuel are deducted from the crew's shares of gross earnings; the result is that, if the trawl comes up with little or no catch, the comparatively inexperienced crew will decide to anchor, rather than risk another tow or expend more fuel in shifting ground[15].

Devising and applying appropriate systems of incentives and sharing of costs and earnings is part of the art of management of deep-sea fishing fleets. Another is to assure the master-fisherman that his vessel, fishing gear and equipment are the best available, both in order to maintain the self-confidence that is so necessary a part of his make-up, and also to eliminate a whole set of excuses for a poor catch. One state fishing corporation observed by the author operated the normal civil service procedure of inviting tenders for supply of fishing gear, stores, maintenance and repairs, and choosing the cheapest; by so doing they incurred losses of earnings far greater than the amounts saved, because of poor performance and loss of fishing time caused by poor equipment and consequent low morale.

The task of achieving maximum production, or maximum profitability, from a fleet of ships operating in unpredictable circumstances, many miles out of sight, is one which requires special knowledge, special experience and a special approach. One of the reasons why the introduction of offshore fishing fleets in developing countries has not always been as successful as predicted, has been the failure on the part of those in charge, and their technical advisers, to recognize the need to make available, or to acquire, the special skills and arts of the successful managers of deep-sea fishing fleets. Sometimes there may even have been insufficient thought as to whether, and if so how, these skills and arts can be effectively exercised in the context of the local social attitudes and customs, and the local style and practice of business management.

Fisheries planning

One of the advantages of the fisheries is that the unit system of production and distribution is usually very small, except in those cases requiring large harbour works, major new roads and the like. Even if the sustainable yield of a new fishery is estimated to be very large, there is seldom the need to decide on building a fleet of perhaps hundreds of vessels straight away: five or ten can be built, experience gained, and the technical specifications and methods of operation adjusted, before building perhaps another twenty or thirty vessels, and so on. The final size of the fleet can then, with luck, be adjusted to achieve something near the optimum balance between catch and effort. Surprisingly, it is only very recently that this advantage has been recognized and acted upon[33], partly no doubt because governments and international aid agencies tend to prefer big projects, and to be impatient of what they regard as delay. This in turn is partly because legislators and their advisers tend to assume that estimates of long-term sustainable yield, based on resource surveys, are reliable and accurate predictions of future landings.

The skills of the fisherman

The skill of the hunter is one of a number of inter-related skills and abilities that the fisherman has to possess and exercise. He has to take a boat or small ship out into the open sea, often in bad weather, carry out fishing operations – which will require the hatches to be open from time to time – and to bring the ship, crew and catch safely back to port. The crew have to be able to handle fishing gear and catch on a platform that is often awash and in violent motion. The masterfisherman has to find and capture wild animals that he cannot see, in vast and largely barren areas of the ocean, and over the abundance and movements of which he has no direct control. In the course of these operations he may have to decide when is the best time to stop fishing and make towards the market.

How to make the best use of echosounders, warp tension meters and other fishing aids was discovered, not by the engineers who developed them, but by the highly-successful skippers who were the first to be persuaded to try them out. A good fisherman has the ability to discern the useful information from his fishing aids despite a great deal of 'noise' injected by ship motion, vibration, electrical interference and so on. He has to collate and analyse information coming from the radio, the radar, the sonar or echo-sounder, the warp tension meters, the fishing gear itself, and from the catch, also that from his own memory and the experience acquired by his predecessors. On the basis of this analysis, which may be a partly subconscious process, he

Plate 11 The life of the fisherman: outward bound in heavy weather in the North Atlantic: a view from the wheelhouse (*Photo: Torry Research Station*)

decides what is to be done next. He continues to do this, to monitor the handling of gear and catch, and to exercise good seamanship, for many hours on end and perhaps for several days. The same concentration and dedication typifies the successful fisherman in the Third World.

Leaving aside the fact that many people become too seasick to perform effectively in a small ship in a seaway, it seems likely from the foregoing that fishing must have become a specialized activity almost as soon as sea-going craft were invented. The part-time fisherman seems to contribute only a very small proportion of total world supplies. Even those able and willing to go to sea in fishing vessels seem to become more reluctant to do so as the society to which they belong becomes more affluent. In the long run, the effective constraint on fish supplies in some parts of the world may not be the long-term sustainable yield of the resources, but the availability of men able and willing to harvest them. If, therefore, there is a possibility that any particular part of the world may have to rely, in future, on the fishery resources it possesses, as a significant source of food or wealth, it would do well to ensure, meantime, that the uncommon knowledge, skills and abilities that would be required, are developed or preserved. They would be difficult and expensive to develop again if ever they were lost by the advanced fishing nations. The performance of the nations that have only recently taken up deep sea fishing suggests that some, but not all, communities have a natural flair for it; others may possibly have difficulty in developing once again the skills, attitudes and psychology of the hunter.

Plate 12 In the wheelhouse: an array of navigational and fishing aids (*Photo: Sea Fish Industry Authority*)

Training

The skills of the fisherman are essentially practical skills, which are learnt for the most part at sea, in the course of commercial fishing voyages, supplemented perhaps by courses of instruction on fishing simulators and other practical training aids. These skills are learnt, but it does not necessarily follow that they can be deliberately and formally taught. Nautical colleges play an essential part in the education and training of the modern fisherman: they teach him navigation, safety at sea, how to run a ship, keep accounts and records; they teach him some biology and oceanography, the principles and practice of the care of the catch, and something about fisheries, fishing vessels and methods in other parts of the world. Schools and training vessels can inculcate the practical skills of the sailor, the manual skills of the fisherman, and the knowledge and skills needed by the seagoing engineer and radio operator. None of this will teach a man to be a successful fisherman.

Nor is it possible, as yet, to identify and select the potentially skilful fishermen during their schooldays. If, therefore, the educational authorities arbitrarily stipulate that certain academic qualifications must be possessed by entrants to training courses for masterfishermen, they may, for all we know, be eliminating the naturally-gifted ones.

There is much confusion and misunderstanding about the proper roles of educational establishments, training programmes

Plate 13 This Korean trainee can be taught the skills of net mending, but whether he will ever become a successful fishing captain is a different kind of question (*Photo: FAO*)

and training vessels, on the part of international agencies that provide technical assistance and development aid, and on the part of the recipient governments. Too close an analogy is drawn with agriculture and with other occupations. Some schools of fisheries are more suited to producing administrators and technicians than to producing the fishermen, fleet managers and managers of shore plants that are needed by the industry, and the task of training these latter categories of people is sometimes neglected.

Decision-making

The exercise of his practical skills and abilities calls for a high degree of self-confidence and self-reliance on the part of the masterfisherman. This may account, at least in part, for the frequent failure, in the field of fisheries, of that other institution so actively fostered by international agencies concerned with technical assistance: the cooperative. There are successful fishermen's cooperatives, some of them long-established; study is required to ascertain why, and in what circumstances, these exceptions are successful.

The masterfisherman must be capable of making up his mind quickly, and arriving at reasonably good decisions, or else he and his crew literally would not survive very long unless they were extraordinarily lucky. This may be the reason why innovations are implemented quickly in the family-owned fleets, given sufficient financial resources or assistance – often more quickly than in the large fishing company, with its longer decision-making chain, and where those in control of financial resources are more remote from actual fishing operations.

Contrary to general expectation, the small-scale fishermen of the third world are by no means always fundamentally opposed to technical change; apparent lack of initiative or backwardness can be due to a lack of awareness of what is possible or available[34], or to a lack of the financial or other reserves that are required before it is possible to accept the risk of trying something new.

The self-reliance and self-confidence of the skipper-owner make it especially difficult for those concerned with fisheries management and development to influence his investment decisions. The fisherman who has come to the conclusion that one more new boat, for himself or a close associate – or a bigger or more powerful boat – would be a good investment, is naturally impatient of those administering systems of licensing or financial assistance if he meets with a refusal. To conduct a serious discussion with a skipper-owner about whether it is in the interest of the community as a whole that he should build an additional boat, or a boat to replace an existing one, and if so, of what size and power, requires a very thorough and detailed knowledge of the fisheries, the fishing operations, the market and the technology. Yet it is not enough to leave these matters to be settled entirely by the operation of market forces, for reasons put forward elsewhere in this book, including for example the natural tendency to build new boats of ever-increasing size and power, and for other reasons which will now be discussed.

Range of individual performance

Everybody is aware that some individual skippers and vessels perform better than others, but the range of variation surprises many. Where the landings statistics relating to particular fisheries or fleets have been analysed, it has been found that the annual landings of individual vessels can in some cases vary over a range of as much as three to one (in terms of weight of fish landed). It is not entirely a matter of skill: the best skippers tend to attract the best crews and command the more powerful and better-equipped vessels; they also tend to spend a higher number of days at sea in the year. Over shorter periods, the range of performance will be wider still.

To the extent that costs, especially crew costs, are related to gross earnings and hence to size of catch, and since in a given fishery the biggest catches tend to be taken by the newer vessels, which are likely to be bigger, more powerful and more costly to run, the range of variation in profitability is not likely to be as wide as the range of individual annual landings. (There are however remarkable differences in fuel consumption between virtual sister-ships engaged in the same fishery, due presumably to the personalities and styles of the skippers.) Nevertheless, when the fleet as a whole is enjoying a modest profit, on average, there is likely to be a very substantial and perhaps vocal minority running at a loss; when the fleet on average is making a loss, it will still be possible to cite a substantial number of boats and fishermen that are doing quite well; if no vessel is running at a loss, some boats will be very profitable indeed; when (as in the classic case of free entry to a common resource) average profitability tends to zero, there will be a fair number of boats making far from negligible losses and a roughly equal number making modest profits. In such a situation it can be very difficult for the legislator or administrator subject to lobbying to discern what is the real financial position of the fleet, unless he has before him sufficient information to indicate the range of variation.

The classic curve of catch *versus* effort, upon which depends so much of the theoretical argument on fisheries management, as shown in *Fig 1*, deals in aggregated statistics. What the fisherman tends to see is *Fig 2,* which presents the same information as *Fig 1*, except that the catches have been divided up between the individual vessels. The newer vessels would be represented mostly by points in the upper half of the diagram and the older vessels mostly by points in the lower half; individual vessels – and individual skippers – will follow various shapes of trajectory over the years, within the area covered by the points, and the impression will be that individual effort, the right kind of boat and the vagaries of fortune are the dominant factors, and not the relationship between effort and catch, unless and until the significance of the downward trend of the points at the right-hand-side of the diagram becomes evident to a number of the more successful of the

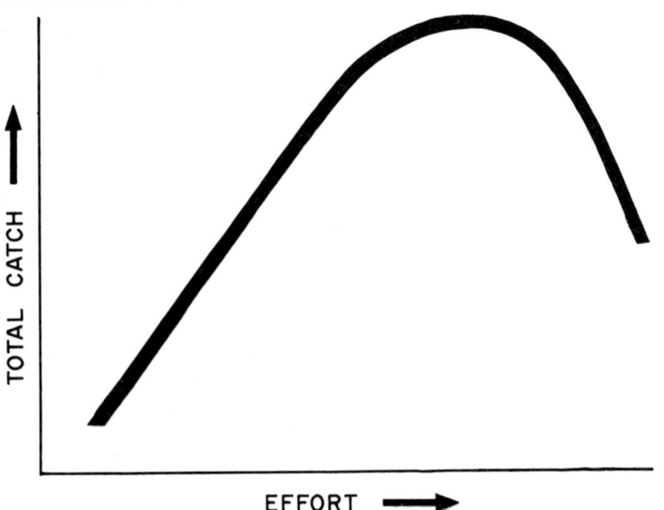

Fig 1 A typical curve of catch *versus* effort as usually presented, the statistics being aggregated for the entire fleet of boats engaged in the fishery

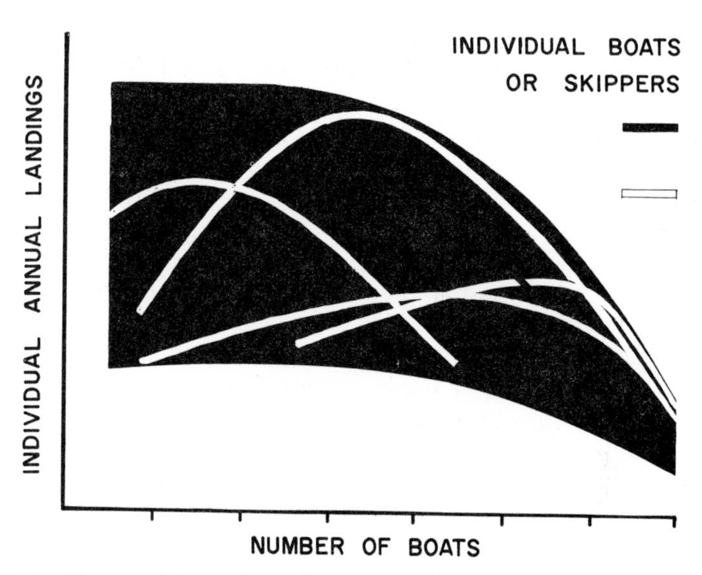

Fig 2 The same information as *Fig 1* but with the catch divided up among the individual boats (intrinsic catching power per boat assumed constant)

skippers. If, as landings decrease, prices rise, this realization may be delayed (*Fig 7*); the constraint may be the prices of competing foodstuffs: the development of the broiler chicken may help to conserve local fisheries, but perhaps at the expense of other stocks exploited in order to produce fish meal for feeding the chickens!

Figure 2 assumes that fishing effort can be measured by the number of boats. If, in the early days of a fishery, there is only a small fleet, the designs of vessels and gear have not yet been optimized, or if strategic fishing information is still scarce, or if there are too few vessels deployed on the grounds to provide adequate tactical fishing information, then the diagram would look more like *Fig 3*. If, moreover, there is more or less continuous technical development leading to improvements in fishing performance or an increase in the number of fishing hours in a year, we might get something like *Fig 4*, which suggests that if the fishery is being prosecuted near the sustainable yield, it is desirable that there should be very close cooperation between those responsible for fish stock management, those concerned with the fleet building programme and those concerned with technical development.

There is a tendency, at least on the part of the layman, to assume that in the classic case illustrated by the curve of catch

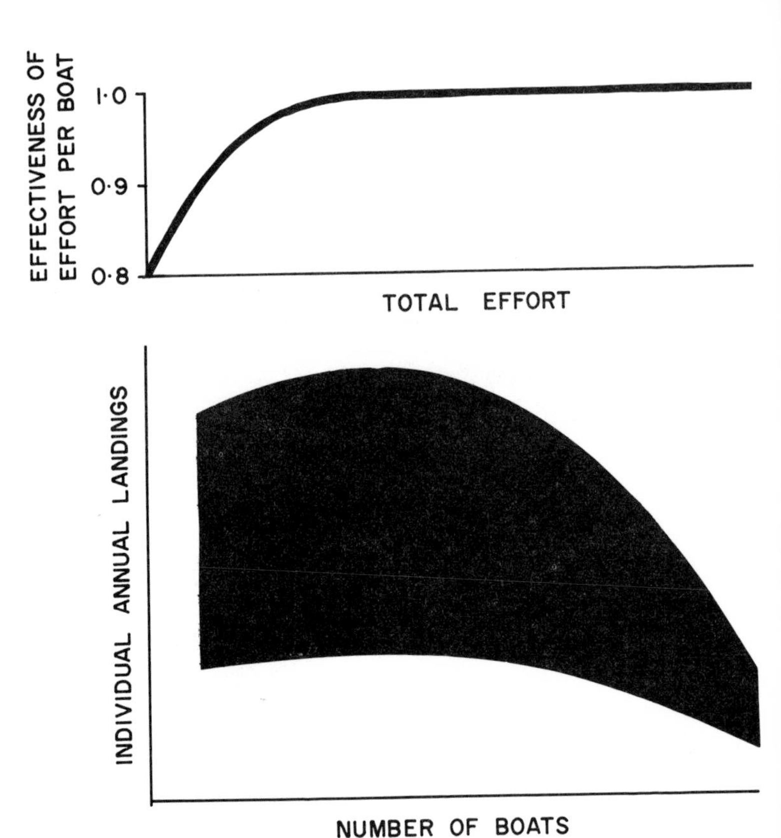

Fig 3 *Figure 2* would be modified as shown if in the early days of the fishery catching power per boat increased with experience and then levelled off

versus effort (*Fig 1*), effort is increasing steadily with time, so that the curve of annual catch against time is the solid line in the lower half of *Fig 5* – *Fig 1* reproduced, but with time along the horizontal axis instead of effort. In one or two developing countries there has been a very rapid build-up of effort in virtually virgin fisheries, as illustrated by the dotted line in the upper half of *Fig 5*; the corresponding curve of catch against time is shown by the dotted curve in the lower half of *Fig 5*. Another case, in which there is a rapid increase in effort as the result of innovation, in a fishery already approaching the sustainable yield – which may approximate to the situation created by the introduction of large purse-seiners into the Northeast Atlantic herring fishery – is shown in *Fig 6*. Whether *Figs 5* and *6* reflect the real responses of any particu-

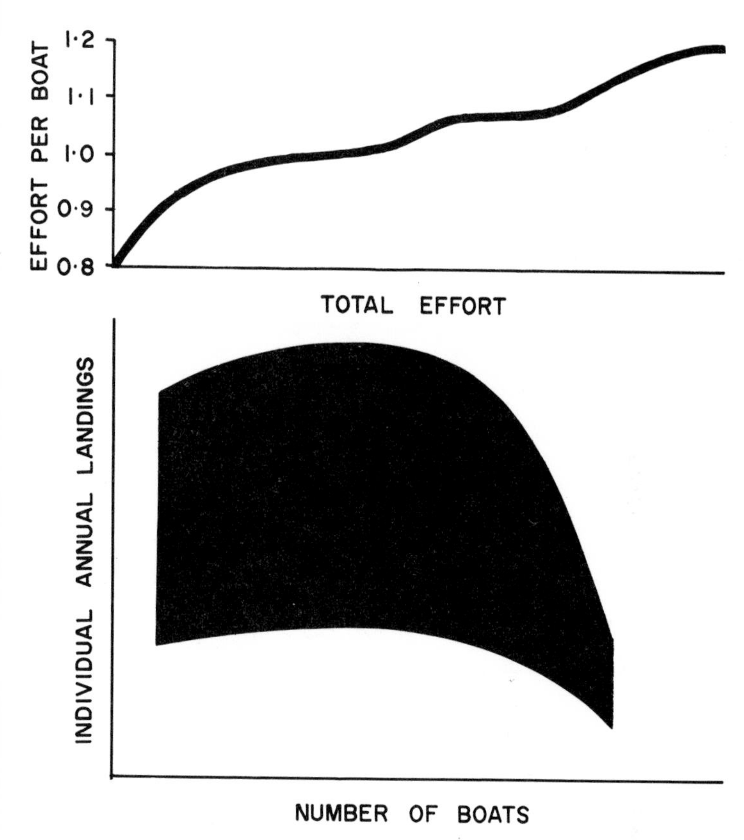

Fig 4 If, through technical development or other means, the catching power continually improves, the picture would be as shown

lar stocks of fish depends on biology, behaviour and the time-scale, but they do suggest that decisions may sometimes have to be taken more quickly than might be inferred by the layman contemplating the classic curve of catch *versus* effort.

If it is decided to reduce effort, it might seem desirable to do so by eliminating the chronically poor performers, so that the fishery could be prosecuted by the least number of boats and men (this would not, necessarily, be the least-cost solution, depending on whether or not some good performers incur higher-than-necessary costs). However, sons do not always inherit their fathers' skills as fishermen, so that the range of performance of the succeeding generation would tend to be as wide as that of the fleet

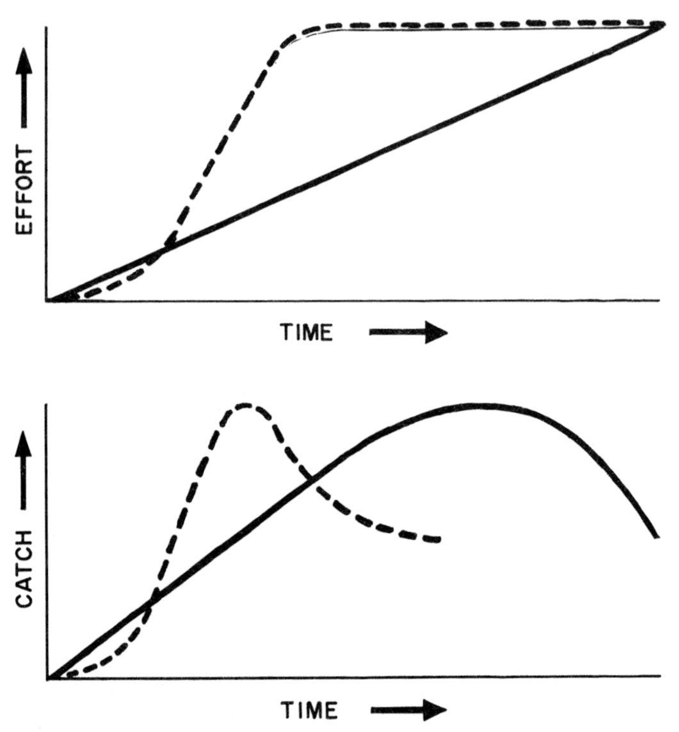

Fig 5 If effort increases steadily with time, the classic curve of *Fig 1* appears as the solid line. If effort increases with time as in the dotted curve, the catch will decrease much more suddenly

before the poorer performers were eliminated. The landings would then fall lower than desired unless allowance had been made for this possible effect.

It is possible, although not proven, that the range of performance might be narrowed by appropriate training, say for the sake of argument from three-to-one, to three-to-two. This would mean, very roughly, an improvement in average performance, and hence a change in effective effort, of about 25%. This assumes that we could discern what training was needed and how it could be done. Little has been attempted along such lines.

The classic route towards improved performance has been more effective technology. Technical development is discussed later, but it is appropriate to look at one particular aspect now: to attain better fishing performance through technical development often means more, or more sophisticated, equipment, and vessels

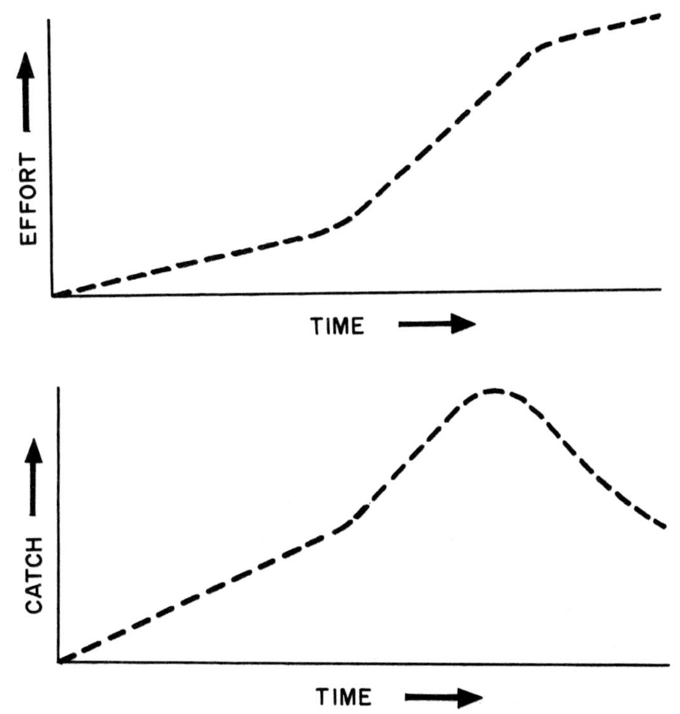

Fig 6 Introduction of new technology may cause a sudden steep rise in effective effort and perhaps a correspondingly rapid decline in catches

that are costlier to buy and more expensive to run. (Technical development and operations research can also reduce costs, and can indicate potential improvements in productivity in other ways, as for example, multiple crewing; or transfer of catches at sea). But, to return to the present point, in cases where it has been shown possible to achieve better fishing performance by fitting additional or more sophisticated equipment, only the more skilled among the skippers may be able to use the new equipment to good advantage, and to derive benefits that outweigh the extra costs; only they may be able to afford to fit the new equipment in the first place. In these circumstances there may be a tendency for the range of variation of performance to become even wider.

New vessels are usually commanded by skippers of higher than average skill because often only they can earn enough to cover not only operating costs but also depreciation on capital and the interest on any loan. In the case of family-owned boats, it is the better skippers who build new vessels. As vessels become more expensive in real terms, they must earn a correspondingly greater amount in a given time if the payback period is not to be extended. If he cannot improve his earnings to this extent, then the original skipper may have to continue in the vessel longer, before it will be possible to hand her on to a less skilled man and to build her successor. The danger may remain that even after the capital depreciation has been earned, and any loan paid off, by the original skipper, the running costs of a big, powerful and lavishly-equipped boat may still be too much for the skippers who acquire her second-hand and third-hand. If some vessels in the fleet are much bigger, more powerful and more sophisticated than others, then the former vessels may have to be retained under the command of the original skippers for a longer time than in a situation where there are only few and small differences in specification throughout the fleet. Thus, unless the range of variation of performance can be reduced, by methods of training yet to be developed, the problem may soon have to be faced of designing and building fishing vessels in such a way that, after a few years of service in the hands of a very skilled skipper, they can be quickly and easily modified to make them simpler and to reduce their running costs. Alternatively, or in addition, a proportion of new vessels could be designed and built not to to suit the skills of the best fishermen, but to a more modest specification which will suit the men of average ability who will acquire them second-hand. This might possibly in some circumstances result in reduced landings – but not necessarily in correspondingly reduced profits – for the very skilled skippers. However, when specifying new vessels, the desirability of building in some reserves of capacity and power, in case of possible re-deployment in another fishery, should be borne in mind.

Objectives and expectations

The earnings required to cover costs vary from fisherman to fisherman, vessel to vessel, and fleet to fleet; so do expectations. What may be overfishing for a community with big expectations and a high standard of living may be a satisfactory enough situation for a community with fewer and simpler needs. The rates of catch required by a young and vigorous skipper trying to earn the capital for a new boat, or to pay off depreciation and interest on loans, are much higher than will satisfy an ageing man who has no intention of replacing his vessel – a vessel that moreover is probably less costly to run than a new one would be. Underlying such situations is the uncomfortable fact that, where fishing is for profit rather than for food, the objectives of the fisherman are not necessarily quite the same, at least in the short term, as the objectives of those who discuss fisheries management in terms of biomass, yield and catch. (*Fig 7*).

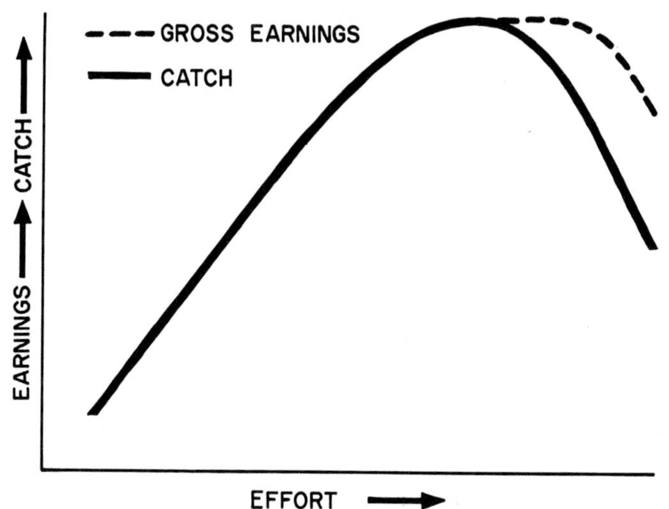

Fig 7 Earnings may not decline as rapidly as catches in a situation where there is overfishing because of compensation through higher prices

One example of divergent aims arises from the fisherman's attitude that if fish is available to capture, it should be caught, not left for another day, or until next year, when the yield might be greater. The motives underlying this attitude are strong: it is not only that the fish may disappear, or that somebody else may catch them, but also that the vessel must be filled with fish as soon as possible in order to allow a quick return home. While there are still some coastal fishermen who set forth daily to the fishing grounds and return home according to the movement of the sun, wind or tides, those venturing further, on voyages of a few days or more, aim to take a satisfactory catch in the shortest possible time in order to bring forward the hour of return to port. (This strong desire to get on with the job, and the willingness to fish hard and for long hours, is not always understood by well-meaning people who have little experience of conditions at sea on board fishing vessels, but are mightily concerned about the number of hours worked in a day. In passing it can be noted that when records were made on distant-water trawlers – for quite other purposes – of the times taken to handle the fishing gear and to gut and stow the catch, the performances of the crews varied very little between the beginning and end of even the longest fishing trips).

Given the harsh environmental conditions and the physical danger, and the difficulty that can often be encountered in finding paying quantities of fish, or even any fish at all, the desire to take as much as possible, while it is available, is completely understandable. One result, however, can be the occasional saturation of the market; sometimes, furthermore, large quantities of this fish may be of low value either because of species, or because of the small size of individual specimens; large quantities of small specimens are sometimes easy to catch, in so far as they are found comparatively near the coast or in comparatively sheltered areas of sea. In such circumstances, and where a system of catch quotas is in operation, the low prices may lead to demands that the quotas should be increased. Overfishing then becomes a possibility.

The apparent exceptions may be some of the bolder and more successful skippers who have confidence in their own ability always to catch sufficient quantities of the larger and higher-valued specimens; indeed, in the North Sea, some of the bigger and more powerful Scottish vessels deliberately seek to land catches of higher-than-average unit value in order to ensure that earnings are satisfactory, bearing in mind the higher costs of bigger and more powerful boats. Some skippers in the Turkish Black Sea trawler fleet contrive to take catches of high unit value by towing nets with mesh sizes much above the recommended minimum – which incidentally allows a vessel of modest power to tow a comparatively big trawl. In other instances, however, landings of high unit value are produced through using the legal minimum mesh and discarding the smaller specimens in the catch. If, then, the rate of discards is underestimated by those concerned with the management of the fishery, there may again be a danger of overfishing. If the discard rate is estimated by those managing the fishery at, say, 20%, and the actual average rate is 25%, this would be within the likely errors in any case; but if the estimate is 20% and the actual discard rate 70% (a figure that has been quoted as being attained on occasion by a practising skipper) then catches are nearly 2.7 times greater than assumed by those managing the fishery.

Consideration of these and many other difficulties in setting total allowable catches and quotas and controlling landings has led Troadec[27] and many others to conclude that the most practicable and effective method of managing a fishery is by the regulation of effort[35].

It is now clear that it was too optimistic to suppose, as was done in some quarters, mainly in the United States, that the prosperity of the fishing fleets will be assured merely by the enforcement of 200-mile limits, regulating the numbers of vessels in the fishery, and leaving the rest to market forces. This very desirable objective might be closer to being attained if, in addition to these measures, the range of variation of performance between vessel and vessel could be reduced; or if it were acceptable to prosecute the fishery at only a fraction of the long-term sustainable yield, so that even the poorer performers enjoyed high enough catch rates to ensure profitability; or if the real price of fish could be raised so as to produce the same result; or if some kind of scheme of equalization of earnings could be worked out. Even then, the vessels must be the right vessels, or the consumer will be paying more than he need pay; also, men must be willing to take them to sea.

The history of the industry shows that solutions of the various kinds mentioned have already been tried, somewhere, some time, but usually only one or two of a number of necessary measures have been applied in any particular instance. A working party on fisheries management set up under FAO's Advisory Committee on Marine Resources Research has reviewed the state of the art[36]. However, as yet, there has not been a sustained, coordinated attack on the problem involving biologists, economists, lawyers, food technologists, marketing men, engineers and all the other disciplines whose activities and potential contributions are touched upon in this book, not to mention the fishermen themselves and those concerned with distributing the products to the consumer.

One of the characteristics of the fisheries industry is the number of different kinds of expertise involved. This is especially true in the field of research and development, to which we will now turn.

4
Technical development in the fisheries

Fisheries development: a definition

To the engineer, the word 'development' means producing new or improved equipment or methods: part of the process of technical innovation. Sometimes the word 'development' is used as a euphemism for restrictive practices; sometimes it means the creation of new institutions, systems or physical structures, whether or not these perform a useful social or economic function. Fisheries development is here intended to mean the processes whereby the exploitation of fishery resources, with appropriate safeguards as, for instance, regarding the yield capacity of the resources, can be made to confer greater benefits than hitherto, or the same benefits at lower cost, the benefits accruing not necessarily always to the fishing community, but to some or other section or sections of society as a whole. Fisheries development can be achieved, in appropriate circumstances, by technical innovation.

Dependence upon distant manufacturers

The basic elements of fisheries technology originated in the very distant past. Although fishing itself must have quickly become a specialized occupation, the skills needed to construct and maintain such craft as the dugout huri and jangada raft, and to make and repair the fishing gear, can be found in artisanal fishing villages, although sometimes the materials of construction come from surprisingly far away. For obvious reasons, the building of planked and decked wooden ships is more specialized but still ancient and fairly local. With the introduction of mechanical propulsion and the construction of hulls in iron or steel, dependence upon distant suppliers and engineering firms increased markedly. The owner of a modern fishing vessel is dependent upon a large number of more or less remote specialist manufacturers, some

Plate 14 Locally constructed traditional fishing craft:

Plate 14(a) Huri on a Somalian beach (*Photo: FAO*)

Plate 14(b) Dugout canoes, Kerala (*Photo: FAO*)

Plate 15(a) Boatbuilding in Brazil (*Photo FAO*)

Plate 15(b) Boatbuilding in Morocco
These scenes would be much the same in Scotland, Turkey, Oman and many other parts of the world (*Photo: FAO*)

probably in foreign countries, and for many of whom the fishing industry represents only a very small part of their total market.

Response to external change

Up to about the middle of the twentieth century, technical development in the fisheries must have been for the most part a slow, gradual and conservative process – practical, step-by-step modification of vessel, gear or methods, each step usually being small, and subject to much discussion, comment, trial and observation, before general adoption. This process appears to the practical man to involve least risk, but it is likely to exhibit what may be described as technical parochialism; although it may be going too far to accept the proverb that 'the self-styled practical man is he who insists on perpetuating the mistakes of his ancestors', there may be an inability to discern new opportunities that may arise out of discoveries and developments in apparently unrelated fields. Moreover, it is limited in the speed and in the degree of response that can be made to external change.

When change became virtually the normal state of affairs – changes in the availability of resources, in the real costs of inputs, in the availability of seagoing manpower, in the markets – a capability for a more rapid and more drastic response to change became desirable. Part of this response is likely to be the development of new or improved methods, equipment and systems. The introduction of new methods or equipment is itself likely to trigger off a series of consequential changes.

Technical development: its changing nature

For about a hundred years, beginning in the middle of the nineteenth century, technical progress was made mainly by more-or-less straightforward adoption of various existing inventions, as for instance the steam engine, the wireless telegraph and radio telephone, the echosounder for measuring depth of water; however, the otter trawl and the Vigneron-Dahl modification should not be forgotten. A little later, somewhat greater difficulty was encountered in making successful and reasonably troublefree use of medium-speed and high-speed diesel engines, because these had less flexibility than the reciprocating steam engine; the manufacturers appreciated neither the severity nor the complexity of the fishing vessel application; neither the shipbuilders nor the vessel owners were able to provide enough quantitative information on duty cycles, loads, speeds and so on, for either main propulsion or trawl winch. The same was true of hydraulic transmissions for deck machinery, despite the fact that in the early 1920s Professor Robertson of Bristol University was already measuring the performance of steam-driven trawl winches by taking indicator cards on the open, often flooded deck[37]. Instrumentation sufficiently rugged and reliable, but at the same time sensitive, responsive and accurate enough to produce useful results in operational conditions on board commercial fishing vessels, was not available until the early 1960s, when Bennett completed the task begun by Professor Robertson, on a winter trip in a stern trawler to the Davis Strait, using instruments and techniques of measurement developed in the aircraft industry[38].

The development of methods of freezing and low temperature storage of fisheries products so as to yield results of high quality took two or three decades, and is noteworthy for the close collaboration of engineers, biochemists and microbiologists, and for cooperation between refrigerating engineering firms and research organizations in the public sector[39].

Increasingly, during the last thirty years, it has become necessary to expend effort to adapt machinery and equipment to meet the special requirements of the fisheries industry, and to invent and develop new specialized equipment for which only the fisheries industry itself has a use: new kinds of fishing gear; deck machinery; fish-finding echosounders and sonars, and other fishing aids; freezing plant and chilling plant; fish-working machine tools, and so on; new fisheries products; training aids and simulators; the application of the methods of operations research to the analysis and improvement of systems, and the provision of management services. Some innovations, like the power-block and the Shetland-type gutting machine, were invented by individuals who were neither professional engineers nor engaged in full-time fisheries development – in at least one case the inventor did not belong to the fishery industry at all – but more and more it has become necessary to engage in deliberate programmes of technical development.

Industry and technical development

The modern fishery industry embraces such a wide range of

sciences and technologies that few, if any, individual enterprises in it can hope to be self-sufficient in the sense of commanding any and all of the various kinds of expertise that may be required in order to make an adequate response to a change in circumstances, or to achieve and maintain a technical lead over their commercial competitors. Perhaps only the monolithic and very large Russian state fisheries organization, and possibly one or two very large companies elsewhere, would be able to make such a claim. The least capable enterprises, in this respect, are the smaller fishing enterprises – the very same family-owned boats and family-owned fishing companies that tend to be the quickest to implement successful innovation and to be the most successful enterprises in financial terms.

The shipbuilders and manufacturers of machinery and equipment are seldom in a position to do enough. Not very much marine equipment is mass-produced; its development is unusually costly and time-consuming, requiring as it does that scarce and valuable technical staff spend a good deal of time at sea, much of it unproductively; many designs and innovations cannot be protected by patent, so that if a firm does embark upon a development project, it will suffer all the teething troubles and infelicities of the prototype, and incur all the expense, while competitors share the benefit of its experience; the costs of development cannot often be spread over a long production run.

There have been exceptions. Rudolf Baader of Lubeck was a manufacturer of metal-working machine tools who became interested in the requirement for a mechanical means of splitting herring, and developed a successful machine. This is a field where success may depend more upon flair than upon inputs of money and manpower. Subsequently, at a time when German government policy encouraged industry to spend money on research and development, the development of a mechanical filleter was brought to the stage where a practicable machine was available, just at the time when the demand was beginning to grow. This demand was stimulated by the development of the factory stern trawler, begun by Burney and brought to fruition by Salvesen and Co, a firm of shipowners with a merchant fleet and a whaling fleet; they availed themselves of support and advice from firms of shipbuilders, trawler owners and refrigerating engineers, and from the Torry Research Station in Aberdeen[40]. At about the same time, in the United States of America, the land of private enterprise, the power-block was invented by Puretic, who was a fisherman, and developed and marketed by an engineering firm. The introduction of diesel-electric propulsion for big trawlers owed much to the readiness of the Siemens, AEG, Metropolitan-Vickers and English Electric companies to send senior engineers to sea in commercial trawlers to study the possible application of such systems. The firms of Kelvin Hughes and Simrad established and kept their technical leads in fish detection, by echosounder and sonar, largely by maintaining a small team of engineers who went to sea in commercial fishing vessels. But these are the exceptions that stay in the memory.

One manufacturer whose engines were for long popular in the Scottish fleet once remarked that there were more of his engines in the public transport system of one large city than in the whole of the fishing fleets. Often the fishing industry is dependent upon special engineering products manufactured by large firms. The aluminium extrusions used in plate contact freezers were first

Plate 16 Like their counterparts elsewhere in the world, these Senegalese fishermen are now dependent upon distant engineering manufacturers (*Photo: FAO*)

made by an aero-space company, the original application being for watercooling the flight decks of aircraft carriers operating jet-propelled aircraft. In this case, the freezer application is now the dominant one, but in this and many other instances the fishery industry represents only a small fraction of the demand for the manufacturer's products: he may be of vital importance to the fisherman, but the fisherman may be of little importance to the manufacturing firm's accountants and shareholders.

In any case, the main benefits of any improvement in methods or equipment introduced into the maritime field accrue, not to the manufacturer, but to the user. The responsibility for ensuring that adequate programmes of technical development are formulated and executed must therefore lie with the user. If he is not capable of doing this, some other organization must do it on his behalf – usually an organization in the public sector.

And so, for example, practical gutting machines were the result of cooperation between public sector development agencies, private inventors and manufacturers; the modern freezer trawler and the single-boat midwater trawl were developed as a result of initiatives and pilot projects in the public sector, involving collaboration between government research establishments, fishing enterprises and manufacturers. Many other widespread innovations have sprung from work in the public sector in recent years and this is likely to be increasingly the case as the costs of development work rise.

The role of science

During the last thirty-odd years, various innovations have been introduced as a result of deliberate technical development programmes in such countries as, for example, Canada, the Federal Republic of Germany, Japan, Norway, the United Kingdom and the United States: the single-boat midwater trawl; fish pumps; chilled sea water stowage; the freezer trawler; net drums and power-blocks; warp tension meters and other fishing aids; systems of transfer of catches at sea; and many other changes in methods, equipment and products. The institutional frameworks have varied from time to time and place to place; one common factor has been the involvement of industry – both user and supplier – with the public sector often providing the technical manpower to carry out development work in the field, together

Plate 17 Some of the developments of the last thirty years:

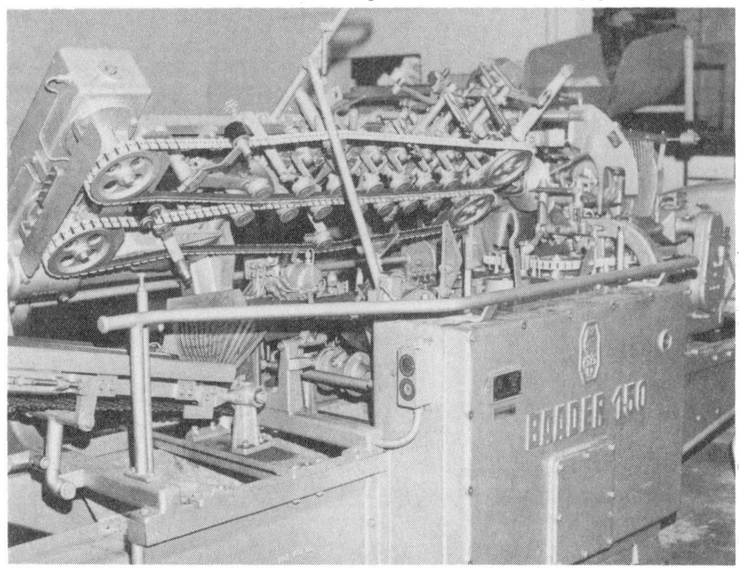

Plate 17(*a*) Fish filleting machine (*Photo: Walter Fussey and Son)*

Plate 17(*b*) Vertical plate freezer *(Photo: Torry Research Station*)

Plate 17(*c*) Hydraulic power-block (*Photo: Sea Fish Industry Authority*)

Plate 17(*d*) Soft fendering systems for transfer of catches at sea (*Photo: Sea Fish Industry Authority*)

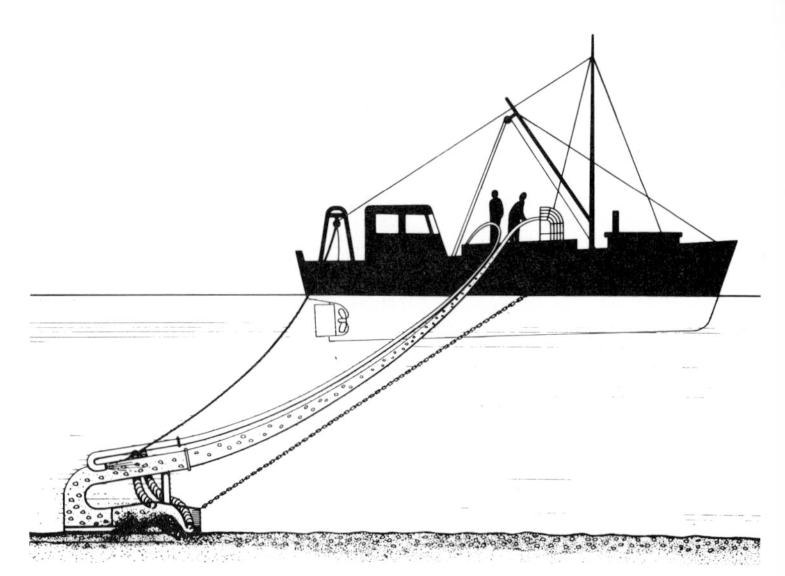

Plate 17(e) Hydraulic mollusc dredge. *(Drawing: Sea Fish Industry Authority)* Similar developments have taken place in the electronics field (see *Plate 12*)

with other inputs, as well as often being the main driving force behind the project.

Most projects have originated in investigations carried out at sea in commercial fishing vessels, on the fish markets at the ports, in commercial processing plants and in transport networks and consumer outlets inland, whereby opportunities were identified for applying new methods and equipment, or for improvements in existing systems. There have been very few instances during this period of a piece of formal, systematic scientific research, in the fields of fisheries and of food science as applied to fish, leading to new technology, but then, in many fields of technology, such occasions are much rarer than the layman, and many scientists, suppose. Research has provided supporting information, defined the limits within which the technical solution has to be found, and has elucidated the reasons why it works (or does not work), rather than leading to technical breakthroughs.

The situation has more often been that the engineer engaged upon a development project has requested a more precise delineation of the constraints within which he should seek a solution, than one where the scientist has produced the germ of a new idea which the engineer has the task of applying to commercial use.

The reason is that, as has been emphasized earlier, fishing is a practical art, using methods and equipment developed in their basic forms in prehistoric times, the workings of which are not even now well understood. One cannot model a complete fishing situation in a laboratory. We can not, as yet, predict accurately the behaviour of a small ship in a wind and in waves of dimensions of the same order as those of the ship itself; the behaviour of models of trawl nets in flume tanks has to be interpreted, not only by the engineer to allow for physical scale effects, but also by the skilled fisherman; we do not yet know how to consult the fish itself. Twenty years ago it became all too apparent that it is not just a simple matter of designing a trawl net with a bigger mouth opening[41].

Even the proper techniques of freezing and cold storage were elucidated by practical trial and error, albeit guided by general scientific principles and good common sense; however, the high degree of perfection to which this and other methods of preservation have been brought is the result of patient investigation leading to a deeper understanding of the underlying physical, chemical and biological processes.

There is good reason to suppose that the contribution of science to fishing technology will be much greater in the future. Until only a few years ago, the most powerful means of observation of fish distribution and behaviour available to the scientist, were those used by the commercial fisherman, and the latter was in a position to acquire more information and to acquire it more quickly than the former. Now there are several more powerful means of observing and measuring what is going on in the depths of the sea, although they are expensive, and the number of scientists doing research of this and allied kinds is smaller than many people suppose, owing to what is seen as the over-riding necessity of providing scientific back-up to systems of fishery management, and to develop those systems so as to make them more effective. Some progress is however being made on the study of various aspects of fish behaviour, including response to stimuli, and to chemical stimuli as well as physical. We may therefore be on the verge of discoveries and developments that will replace the technology that has been in use for so many thousands of years.

We are not yet, however, at the stage where it is reasonable to expect that a comparatively small expenditure on science will produce large economic benefits: the situation is not analogous to that in agriculture, where the development of pesticides and fertilizers, and of new genetic strains of animals and plants, has produced enormous increases in value of output for the expenditure of comparatively very small sums on research. Fishing technology, like most maritime technology, advances with such support as science is capable of providing, but largely by practical trial-and-error on the full scale, with the concomitant risk and expense.

Governments, in their attempts to manage the fisheries, seek scientific advice, and support programmes of scientific investigation; it is natural that the formulation and execution of such programmes should normally be entrusted to oceanographers or fisheries biologists (or to graduates of schools of fisheries, which is usually much the same thing). It is less obvious that the same individuals, or people with a similar education and background of experience, are the best sort of choice to oversee programmes of research and development aimed at improving the performance of the industry in the harvesting, utilization and marketing of fish. Yet this often seems to be assumed. In the light of the relationship between fisheries technology and science outlined above, the assumption is not easy to justify. It seems to derive in some cases from the belief that there exists a profession or group of specialists called 'fisheries experts'. In the kind of lists of specialists that are kept by development banks and other agencies concerned with recruitment of consultants, it is normal to find agricultural experts sub-divided into various kinds: agronomists; experts in soils, irrigation, land use, locust control, animal husbandry, agricultural machinery, rice, tea, soya beans and so on and so on; this long list will be followed by a single entry: 'fisheries'. The result is that a good deal of advice is sought from the wrong kind of expert, especially in the areas of fisheries technology and fisheries development, and the person consulted does not always feel compelled to draw attention to the fact that he is not really the right person to advise on whatever subject has been raised when it is outside his particular experience and field of expertise.

Management of research and development

One consequence of the fragmentation of the fishery industry, its lack of enthusiasm for cooperative ventures and its heavily practical nature, is that it is not accustomed to managing research and development. Care is needed to devise appropriate arrangements for close and frequent consultation regarding the objectives and priorities of R and D programmes, while fending off attempts at detailed direction as to how the objectives should be achieved. This must include an ability to perceive when a technological solution is appropriate and best, and how long it is likely to take to achieve it: for example, when the Scottish trawling industry was faced in 1964 with a glut of small haddocks owing to an exceptionally large year-class of recruits to the fishery, it was almost certainly unrealistic to ask for a crash programme of development to be mounted to produce a mechanical block filleter; other, non-engineering, solutions had to be found. As a matter of fact, a machine was available when the next big year-class entered the fishery, four years later.

Defining operational requirements

Not many years ago a trawler owner said to an engine builder: 'Look what we have done to your engine! You should not have designed your engine in such a way that we were able to do that to it!' All maritime organizations recognize three levels of truth regarding what happens at sea: first, what the designers of equipment, suppliers, shore managers, administrators and legislators believe or assume goes on at sea; second, what the officers and crews of the ships say goes on at sea; third, what actually does happen at sea, which can be ascertained only by going and finding out for one's self.

There is, all too often, a need for technical information that the user is not equipped to provide, and that the supplier is unwilling to go to the trouble of obtaining at his own expense: what exactly the equipment is expected to do, in technical terms that can be understood by a design engineer; how the equipment is likely to be treated; in what sort of environment it has to work. Obtaining such information is one of the essential functions of any organization devoted to technical development in the fisheries. It is costly and difficult, because of the environmental conditions and the need to go to sea.

Use of research vessels

If new methods and equipment are to be well-suited to their pur-

Plate 18 The ability of the industry to respond to changing circumstances requires a thorough understanding of technical problems and possibilities, which can be obtained only by measuring the performance of vessel and equipment in operational conditions. These plates show:

Plate 18(a) Instrumentation on board a commercial fishing vessel (*Photo:* Sea Fish Industry Authority)

Plate 18(b) Measuring the loads on a trawl winch during commercial fishing (*Photo: Sea Fish Industry Authority*)

Plate 18(c) A section of propeller shafting, strain-gauged to measured thrust and torque (*Photo: Sea Fish Industry Authority*)

Plate 18(d) The engineers engaged in such work have to endure the same conditions as the commercial fishermen (*Photo: Torry Research Station*)

pose, they must be designed and developed against a background of adequate knowledge of the conditions of service, which can be acquired only by preliminary observation and measurement in the commercial situation. If the new methods and equipment are to gain acceptance, they must be subjected to trial and subsequently to demonstration in fully operational conditions, on board commercial fishing vessels, or in commercial processing plants or distribution networks on shore, as the case may be. None of the innovations mentioned earlier would have come to much, without the active collaboration of vessel owners, skippers and crews, fish merchants and processors, fish retailers, catering experts and others.

The usefulness of research vessels in the field of technical development is therefore limited. There are cases where the proposed new method or equipment is so novel, or so imperfectly understood, that it would be most unwise to expose the idea to trial in commercial conditions until more knowledge and experience has been gained, enough to convince the owner of some commercial vessel that collaboration in the project will not be a complete waste of time. In these cases, the research vessel and the shore laboratory are test-beds for preliminary trial of the new idea on a pilot scale. Research vessels are also useful for conducting supporting investigations under conditions of greater control than are easily attainable in a commercial situation, and furthermore for producing supplies of well-documented raw material, or intermediate products, for further systematic investigation in the laboratory on shore.

The whole subject of the uses and limitations of fisheries research vessels, how they should be manned, operated and managed, and their cost of operation, is now the subject of study by FAO and UNESCO.

Research vessels are often used for two activities frequently undertaken as part of a programme of fisheries development: first, resource surveys and, second, exploratory fishing. There is often confusion between these two quite distinct activities. Exploratory fishing, or indicative fishing, is aimed at producing information on likely catch rates, species and so on, in order to allow an assessment to be made of whether commercial fishing opportunities exist or not, and if so, what size and type of commercial fishing vessel and fishing gear would be appropriate; it ideally requires the employment of an exceptionally gifted commercial fisherman and crew, with wide and varied experience, assisted by fishing technologists, fish technologists and experts in vessel design and fleet management. A resource survey, on the other hand, is aimed at estimating biomass and sustainable yield, and hence at gaining a first rough estimate of how many commercial vessels it may ultimately be possible or advisable to allow into the fishery; such a survey needs to be conducted by scientists skilled and experienced in such matters. There may be many fisheries scientists who would make good commercial fishing skippers, but actual examples of this metamorphosis are rare.

It is also sometimes assumed that the services of a research vessel will continue to be necessary in order to provide information for management of the commercial fishery, when it develops, by repeated surveys of the resource, in preference to and in place of establishing appropriate systems of collecting and analysing statistics relating to the commercial operations. This is usually an uneconomic use of an expensive facility.

New or improved methods and equipment intended for use by the commercial fleet are likely to be best fitted to their intended purpose if they have been developed and proved on board a vessel commanded by an exceptionally able and highly respected skipper, and they will then find much readier acceptance in the commercial fishing fleet. This, indeed, is true of most information generated by workers in the field of fisheries research and development, and is one of the arguments for doing as much as possible on board commercial vessels in preference to research vessels.

Moreover, when a research worker or development engineer embarks in a commercial fishing vessel with a good skipper and crew, he is almost certain to gather a great deal of information and experience which, although it may not be immediately relevant to the primary purpose of his current mission, will probably be helpful sooner or later, either in his own future work or in that of the organization to which he belongs. Fishermen are invariably pleased when anybody takes the trouble to go to sea with them; they possess information and experience which they too often suppose is already known to scientists and engineers; they will listen the more readily to proposals and findings if the background against which these have been drawn up is what they would regard as realistic.

Although much of the work of technical development will take place in the premises of equipment manufacturers and in the vessels, premises and transport vehicles of the fishery industry itself, those responsible for the execution of the development projects need a base from which to carry out their work. Ideally this should be on or near the quayside, at a fishing port, where the vessels and shore installations will serve as a constant reminder of the purpose of the work.

Research and development in support

The teams of engineers, operations analysts, seagoing technicians and others engaged in programmes of technical development often need the advice and assistance of their colleagues in research establishments, but the kind of support they seek will by no means always be found in fisheries research establishments. Indeed, because the fishery industry makes use of such a wide variety of engineering products, involving several different technologies, those actively engaged in technical development must establish and maintain close contacts with research and development organizations in a number of fields outside fisheries, and some of these contacts will be at least as important as those with the fisheries research establishments proper: for example, establishments engaged in research on ship hydrodynamics and other aspects of the design and operation of ships.

At the same time, those responsible for technical development in the fisheries must realize that the really valuable element of their own particular expertise is their knowledge and experience of the fisheries application. This was the basis upon which, for example, the responsible authorities in the United Kingdom delineated the boundaries between the fields of interest and responsibility of what are now the National Maritime Institute (ship hydrodynamics and related matters), the Ship Research Association (shipbuilding and related matters, ship operation) and the Industrial Development Unit of the Sea Fish Industry Authority. It was never much in doubt, of course, which group would undertake the major share of seagoing in commercial fishing vessels!

The engineer engaged in fisheries development must therefore expect to derive satisfaction in his job mainly from seeing the results of his efforts applied in the commercial fisheries industry, rather than from a reputation as an expert in some widely-recognized branch of engineering, and certainly not from the production of numerous publications in learned journals. That is not to say that the work lacks intellectual interest and challenge: some fisheries applications can be very sophisticated, and break new ground, as for instance the design of freezing plant.

One measure of the usefulness of organizations devoted to applied research and industrial development should be the proportion of staff who are offered appointments in the industry concerned or in allied industries. In fisheries, however, the small size of the average commercial enterprise makes this outcome less likely than one might wish.

Organization of research and development

The world of fisheries research and development is itself often fragmented, institutionally speaking. In the United Kingdom, the government-funded establishment responsible for research and development in the handling, processing and preservation of fish as food, and other fisheries products, is the Torry Research Station. This formerly came under the Department of Scientific and Industrial Research (which no longer exists). Torry was therefore funded at that time under a completely separate vote from the research establishments belonging to the Fisheries Departments. The research work leading to the development of freezing at sea was financed and carried out quite separately from work on fishing gear technology, which was the responsibility of the Fisheries Departments.

In the course of the 1950s, two separate and expensive projects were mounted, both aimed at improving the economics of the distant-water trawler fleet. One was to develop a better trawl that would enable the fleet to enjoy higher average rates of catch; the other was to freeze the catch on board, to allow the fleet to spend a higher proportion of the total time at sea, actually catching fish. The introduction of a new trawl would involve no radically new practices, and would require much less capital investment, and so, in the opinion of many, was the solution to be preferred. If the two development programmes had been managed and financed through one unified, coordinated R and D organization, financed from a single vote, there might have been greater pressure to pursue one option or the other, but not both; if there had been consultation on the question of which project should be continued, the balance of advice would probably have favoured the trawl gear project. In the event, this failed. Both solutions were in fact pursued, both being put to practical trial in commercial conditions under the auspices of yet a third agency, with access to yet other dedicated funds for R and D, the White Fish Authority.

Divided organization may however lead to missed opportunities. Throughout the 1960s, the stocks of cod in the North Atlantic came under increasingly heavy fishing pressure; alarm was expressed by those whose official concern was the continued well-being of the fisheries. Much of the catch was being processed into fish fingers. During this time, there took place under the auspices of the White Fish Authority a very thorough survey of the attitudes of children to fish in school meals, as measured by the wasted food left on the plates. Fish came out very well, provided that it was presented in the form of fish fingers; moreover, provided they were in the form of fish fingers, there was almost no difference in the acceptability of various demersal species, some of which were cheaper and in greater abundance and availability than cod. There seemed to be an opportunity to reduce the fishing pressure on cod. One leading firm in the industry, however, advertised that their fish fingers would continue to be made from cod, and others followed suit.

In Denmark, a system was developed for cooling the catches on board the vessels of the North Sea fleet engaged in supplying raw material to the reduction plants producing fish meal and oil. The use of the system resulted in meal of higher quality, and it was estimated that, averaged over the year, yields would be improved by about 10% or, to produce the same amount of fish meal as before, catches could be reduced by a corresponding amount. Those concerned with fishery management did not appear to regard this as important enough to justify any special action.

More generally, and as pointed out by Iles[42] in regard to the herring fishery of the Gulf of Saint Lawrence, the manner in which the catch is handled, preserved and utilized, has a bearing upon the quantities of raw material that have to be caught in order to provide the fishermen with a given income. Although the literature on fisheries management recognizes the desirability of taking the utilization of the catch into account, in practice there have been few schemes of catch allocation, quotas and the like, which do so, although schemes have recently been proposed whereby the allocation of catches will be such as to give preference to those who will add most value. In former times there were schemes whereby the amounts of fish utilized in different ways were regulated, but these were aimed at maintaining prices at time of landing rather than at resource management. One more example of the interaction between resource management, biology and technology: post-spawning fish are usually in poor condition, and do not freeze and cold-store well; it is therefore doing the consumer a disservice to open the fishery in the immediate post-spawning period.

The fish finger and fish meal are examples of technology that may have at least as profound implications for fishery management as those already cited; another is surimi. In recent years, the food technologists have developed methods of comminuting, mixing, and restructuring fish flesh so as to make a great variety of standardized products, including products resembling, for example, mayonnaise and cheese, as well as fish fingers, sausage, kamaboko, pastes, simulated fish fillets and simulated crustacean tails; the raw material may be taken from a wide variety of fish species. When the processing has to begin on board the catching vessel, as in the case of surimi, it becomes less practicable to monitor the sizes and species of fish that have been caught, unless there is an observer on board.

At the same time, however, the marketing man may be less prone to take refuge behind allegations of what the consumer will, and will not, accept, and be more ready to market whatever species happen to be in greatest abundance: aided by the food technologist he can, if he so chooses, have available a variety of standard manufactured products that may not always be made from the same species. If the consumer continues to accept such products, a further possibility arises: we might exercise the option of catching ten tons of capelin in preference to the equivalent one ton of cod; indeed, we might even still produce some high-value traditional species by feeding the offal remaining after processing the capelin, to captive turbot, sole, salmon or shrimps.

Be that as it may, in the fields of fisheries management and the associated scientific investigations, those responsible have nearly always been constrained, by the resources available for the work, to concentrate upon the study of the fisheries for those species that the market regarded as important at the time. Market preferences, however, change, and can be changed. The food technologist and the marketing man have a contribution to make to the solution of the problems of fisheries management, the debate on which has hitherto been conducted for the most part by biologists, economists and, latterly, lawyers. Earlier, we saw that in future the biological scientist should be able to make significant, perhaps revolutionary, contributions in the field of fishing technology.

Once again, the conclusion seems to be that there is a need for a greater exchange of information and ideas among the many different kinds of experts, and a closer coordination of their efforts. One problem that may then arise is the scarcity of individuals with sufficient variety of experience and breadth of vision to enable them to give due consideration to each part of the picture and to comprehend the whole.

The developing countries and technical development

In the developing countries, much of the progress during the last few decades, and up to now, has been accomplished by more-or-less straightforward transfer of technology, and this is an underlying assumption in the formulation of many programmes of technical assistance. However, existing equipment and techniques are not always well suited to the local circumstances in the developing country, and only the easier opportunities can be handled satisfactorily in this way. There will be a growing need for programmes of technical development to adapt imported technology to local requirements or to develop genuinely local solutions. In these cases, it is even less likely than it is in the advanced fishing nations that the local boatbuilders and suppliers of equipment will be able to formulate and carry out, or support, programmes of investigation at sea in operational conditions on board commercial fishing craft, or in shore-based processing plants; nor will they always possess adequate facilities for design, let alone development. It is possible, however, that the requirements for such expertise might be met through international technical assistance at comparatively low real cost. There are several developed countries where, by reason of the importance of the local fishery industry, or because the local shipbuilders and engineering industry are involved in building fishing vessels and equipment for fishing vessels, it will have been decided to provide from the public sector substantial technical support, embracing a wide range of expertise. In any one fishery, it is not likely that the current problems and opportunities will require the simultaneous deployment of the whole range of sciences and technologies that may be covered; nevertheless, once any individual specialist has become familiar with the peculiarities of the fisheries industry and has become knowledgeable, experienced and useful in his special field, it is desirable that his services continue to be available to the fishery industry whenever required. Thus, a proportion of these experts could well be available at any given time to provide advice and technical assistance in other parts of the world. There is, for example, one pre-eminent expert on systems for stowing fish on board in chilled sea water, whose services have been made available in many parts of the world, and another whose special expertise is in the rigging and use of the single-boat midwater trawl. These examples also illustrate that there are so many different specialisms in fisheries, and so few experts, that duplication of effort and expertise in the technical field is decidedly undesirable and, to an extent, unnecessary.

5
Engineers in government

The earliest functions of government in the field of fisheries were the settling of disputes, by diplomacy or by armed force if necessary, and the exaction of rents. Nowadays the activities that are most often in the public eye are those concerned with conservation of the resources and the allocation of catches. There is also the question of profitability, and Crutchfield[43] has pointed out that the fisheries present, in one form or another, all of the major causes of market-mechanism failure that call for public intervention; Smith[44] considers it to be more accurately described as an instance of incentive failure caused by cultural or institutional inadequacies. Whatever the diagnosis, support, financial and by other means, often has to be provided. As was outlined earlier, the manner and efficiency of utilization of the catches has a bearing on all these questions.

Nowadays the involvement of the government does not stop there. There are regulations regarding food standards and hygiene, safety at sea, manning scales, conditions of work, qualifications of seagoing officers and various other aspects of the industry and its activities, which are not usually regarded as integral parts of fishery management, but which can have a significant influence on the industry's methods, equipment and costs of production. Rothschild is right to remark that 'fishery management needs to comprise all public sector decisions relative to fishing'[31,32].

Goverments, technology and economics

The responsibility for formulating and executing programmes of technical development on behalf of the fisheries industry, and for the associated research, is usually borne to a large extent by governments, irrespective of their political colour or philosophy.

The industry cannot often do much for itself, because it is fragmented; most enterprises are small; different sections have different objectives; there is not a great tradition of cooperative action; there is little experience of managing programmes of research and development. Moreover, the fishing industry, or some part of it, is often in financial straits and unable to pay for research and development; this then becomes part of the support provided by the government, often a valuable part. The shipbuilders and equipment manufacturers cannot be expected to mount adequate programmes of technical development on behalf of the fisheries industry, for reasons outlined earlier. The government, on the other hand, is often involved in any case in scientific investigations related to the task of fisheries management and therefore already possesses some useful facilities and expertise; often, also, governments already possess capabilities for work in the fields of food science and technology.

The government itself may profit from an active programme of research and development, because it is a good way of gaining and maintaining the knowledge and experience that are necessary for the competent execution of the regulatory functions mentioned a little earlier, and for 'reducing the real costs of production, reducing waste and broadening the supply base' as the functions of government in the fisheries have been described [31,32].

Such matters as safety at sea, food standards and hygiene, working conditions, ports and harbours and other infrastructures are more often than not the responsibilities of different ministers from the one directly concerned with the fisheries. These other ministries or departments are therefore in a position to make decisions affecting the prosperity and well-being of the fisheries industry. Every one of these fields is a highly technical one, and the departments of state responsible for them usually possess scientific or technical experts. However, few people outside the fisheries industry are at all familiar with its methods, equipment, practices and attitudes, or with the constraints within which it has to operate, and it is an unfortunate fact that, in the fisheries, the behaviour of many systems, both natural and man-made, is somewhat different from that in other areas of human activity, a fact of which the experts in other departments of state are not always aware.

The kind of situation that can develop is, for example, an

attempt to undertake the task of quality inspection at the fishing ports by veterinarians whose only training has been as inspectors of abbatoirs and the meat of warm-blooded animals (fish flesh is usually free of organisms pathogenic to human beings); or again, the allocation of load-lines, with all the associated regulations, to deep-sea trawlers; and again, the application of scales of manning and pay, qualifications and hours of work, appropriate to large merchant ships.

The topic of safety at sea has received much attention in recent years, and particularly the subject of the transverse stability of fishing vessels. Recently, the technical literature has revealed a strong and growing body of opinion among academics and research workers that the standards promulgated a few years ago by the Inter-governmental Maritime Consultative Organisation (IMCO, now the International Maritime Organisation – IMO) a Special Agency of the United Nations, are not altogether adequate[45]. This had always been the opinion of some fishermen and some traditional boatbuilders; perhaps there were few such people at Torremolinos when the draft standards were discussed; at any rate, their views did not prevail. The draft standards differed markedly from orthodox practice in the design of large merchant ships and warships, being, in some respects, already more stringent, although, as many now suggest, not stringent enough; the technical experts who advised governments on such matters, however, usually had a big-ship background. After publication of the draft standards, there followed a period during which some new fishing vessels were designed by consultant naval architects unfamiliar with fishing vessel practice, and who may have assumed that the IMCO standards were stringent enough; some of these vessels were built by non-traditional builders. Subsequent accidents and losses, including some that were not obviously attributable to inadequate transverse stability, inspired programmes of research and investigation which have led to the conclusion already noted, and which furthermore, ironically enough, suggest that there is scope for a more subtle approach to the treatment of transverse stability of certain classes of merchant vessels and warships.

To ensure that the technical knowledge and experience that are brought to bear, in situations like those described above, are commensurate with departmental responsibilities, requires that the fishery industry, or else the fishery authorities acting on the industry's behalf, should themselves have technical advisers who possess sufficient knowledge and experience to appreciate the implications of decisions and actions by other departments of state. These technical advisers should be of sufficient competence to command high respect in their individual professional fields, and they should have direct access to those in authority.

Preferably, the task of formulating and applying regulations to the fishery industry should be delegated to the fisheries department, and the 'responsible' department should confine itself to stating the objectives to be attained and should exercise supervision on that basis. Such an arrangement would have the very considerable additional benefit of reducing the number of government departments and officials with whom the individual fisherman or fish merchant has to deal.

It is, however, unlikely that the fisheries department will be allowed complete and independent jurisdiction over all matters affecting the well-being and prosperity of the fishery industry, except perhaps in those few countries or provinces where the fisheries are of primary economic, social or political importance. Elsewhere, various kinds of institutional framework have been tried, and certain functions have sometimes been delegated to parastatal agencies[46] and independent research boards, alongside the regular civil service departments or ministries. Judging by the fairly frequent investigations into the organization of fisheries administration and of fisheries R and D, followed in not a few instances by actual reorganization – sometimes twice within a few years – no perfect organizational arrangement exists. This is usually assumed to be because of the comparatively minor importance of the fisheries, even in a majority of the leading fishing nations, coupled with their technical complexity, involving, as they do, activities that may fall within the fields of responsibility of more than one other government department: safety at sea, food standards, ports and harbours, and so on. If, however, we accept that 'fishery management' should comprise all public-sector decisions that affect the fisheries[31,32] – and in particular all decisions that affect costs – then it might appear that the only satisfactory pattern of organization of fisheries administration from the industry's viewpoint is for only one department or agency of state to exercise jurisdiction. Certainly, if more than one department is involved, there is a need for close coordination, especially on scientific and technical matters having a bearing on costs.

There is no closely analogous situation in shore-based industry: in a ship such as a fishing vessel, the design is normally space-limited; any substantial change in design, in practice or equipment reacts on other parts of the vessel and the consequences ramify throughout the ship, affecting design, stability, operations and economics in various more or less subtle ways.

The staffing of fisheries departments

To facilitate communication between the industry and the government, many countries adopt the practice of stationing officials at the main fishing ports. Such a system of communication and consultation is very useful, provided that it is always borne in mind that it needs to be supplemented and complemented by a direct flow of objective and quantitative information from the vessels at sea and also that, since much of the activity and information is highly technical, the implications need to be worked out by experienced specialists; often only the specialists will be able to discern that additional information, or different kinds of information, are required in order to obtain a reasonably accurate picture of present circumstances and trends.

One very simple but quite important example is the incidence and rate of discards of fish at sea, both under-sized and under-prized, one of the kinds of statistics related to population dynamics about which there is much doubt and uncertainty. In former days, in England and Wales, there were special seagoing observers who collected various kinds of routine statistical data on board the vessels of the commercial fleet; nowadays, it may be too expensive to maintain a corps of such people (although trade-offs with costs of operation of research vessels might be possible). However, those engaged in programmes of technical development necessarily spend a good deal of time at sea in commercial vessels, in the course of which they can usually spare the effort to collect a whole variety of information and samples related to fishing operations and general conditions at sea, while at the same time they invariably gain an intimate knowledge of attitudes and opinions.

In an industry like the fishery industry, decisions have to be made on the size and composition of the fishing fleets, location and equipment of fishing harbours and the like, which may well involve technical questions such as method of capture; form of products; size, type and power of vessels, and so forth. If it is agreed that the attempt should be made to decide these matters on a rational basis, rather than on purely political considerations and if, furthermore, the processes of consultation are to be anything more than superficial, then the scientific and technical advisers in the public sector have to be not only highly capable but also very well-informed. The best way of keeping well-informed about a highly technical industry as it exists and operates at present, is to have teams of scientific, professional and technical people active in the field – on board the vessels of the commercial fleet and in the commercial premises and distribution networks on shore. A very good way of keeping well-informed about fresh options, new opportunities and the possible solutions to current problems, is to have an active programme of research and technical development.

Some fifteen years ago or more the Fisheries Departments in the United Kingdom introduced a scheme for providing financial assistance for improvements to existing fishing vessels. It was very soon recognized that the scheme should be couched in terms of objectives, such as for example: improved safety and comfort; better catching performance; preserving the quality of the catch; and so on – and not in terms of detailed physical specifications of what would be approved. What is actually done in each case is left to the technical judgement of an experienced surveyor acting where necessary on the advice of specialist colleagues in the research and development field or in accordance with objectives laid down by counterparts in other departments. Any attempt to restrict the freedom of judgement of the surveyor and the fisherman by means of rigid specifications as to what will be permitted could result in excluding novel but valid solutions some of which may be superior to the orthodox, and also and in any case would result in very voluminous sets of regulations, because of the complexity of fishing vessels and the many different ways in which some of these objectives might be achieved; such regulations would be difficult to keep up-to-date when technology is changing rapidly.

There is a strong case for all government requirements regarding construction and equipment of fishing vessels to be monitored by surveyors who are not only professionally qualified but also knowledgeable and experienced in methods and practices on board fishing vessels, and responsible to the senior professional adviser of the fisheries authorities.

One very effective way in which the professional staff and technical advisers within the fisheries department can keep themselves well-informed and up-to-date, not only on current methods, equipment and practices, but also on problems and opportunities, is for them to be closely associated with the organization responsible for scientific research and for technical development. This presupposes that these organizations are engaged in active programmes of investigation and development on board the vessels of the commercial fleets and in the processing plants and distribution networks on shore, and that they are in reasonably close communication with their counterparts overseas.

Fisheries policy and objectives

The broad objectives of public policy for the fisheries would appear, in principle, easy enough to comprehend. Depending upon local circumstances they may include, for example, conservation of the resources; socio-economic objectives related to the fishing communities or to urbanization; the securing of food supplies for the present, or developing or maintaining a capability for securing food supplies in the future; providing a regular supply of fish to the consumer at a reasonable price; maintaining or improving the national maritime capabilities; perhaps even deriving the maximum economic rent from the fishery, as is assumed in so much of the literature on fisheries management.

It is more difficult to decide how the broad objectives should be achieved, especially in the face of frequent distractions such as crises of glut and scarcity, bad weather, failure of year-classes, changes in the availability and cost of inputs, and fluctuations and changes in the market, over none of which has the industry much control. The difficulties of deciding how to attain the national objectives are in any case great because the fishery industry is so diverse, and it uses a wide range of uncommon techniques and equipment, often in remote places and in rough environments where it is not easy for the outsider to observe what is going on and see how things are done, or to appreciate the reasons for certain attitudes and behaviour, or to discern either the opportunities or the real constraints on any particular proposed course of action.

In Iceland, and perhaps in one or two other parts of the world, the legislators and administrators will probably have some personal experience of life aboard commercial fishing vessels or of working in processing plants, and almost certainly will have close connections with the industry. It takes some years to acquire a reasonably comprehensive, detailed knowledge of the industry, sufficient to be able to discuss the practical implications of policy, the ways in which it could be implemented, and the objections to specific proposals; or to be able to appreciate what fresh options are available and to assess the soundness of particular technical decisions. Elsewhere in the world it is difficult for those responsible for decisions in the public sector, and their advisers, to acquire sufficient background knowledge. The existence in the public sector of a body of professional and technical people engaged in a practical programme of technical investigation and development, and technical inspection and regulation, and hence thoroughly conversant with current conditions at sea and on shore, and able to assess the implications of particular propositions, can be of great help.

Consultation with industry

Consultation with the industry is of course essential. There is, however, a peculiar difficulty in consulting the fishing industry: those most readily available for consultation are either retired from active fishing, and in such cases, have not always been as familiar with current conditions and practices as might have been expected; or else those consulted are themselves one or more removes from actual operations. As noted earlier, the only way to ascertain what actually goes on at sea is to go and find out for one's self.

On those occasions when active fishermen are available for consultation, they may not marshal their facts and present their arguments in quite the way that the senior civil servant would find familiar; moreover, they may tend to assume a wider and deeper knowledge of the subject under discussion, on the part of individual officials, than the latter in fact possess. The presence of experienced professional advisers with practical knowledge of the subject can then help to illuminate the discussion. Also, owing to the wide range of performance of individual fishermen and their widely differing aspirations and requirements, remarked upon earlier, special care has to be taken to ensure that the facts and views put forward are representative. Views can be conditioned by recent and current crises such as gluts and shortages, and in such circumstances the familiar will receive thorough consideration but fresh options and opportunities may not.

Not so many years ago the British fisheries scientific authorities, concerned at the ever-increasing fishing pressure on North Atlantic cod and the likelihood that in future years it might be in short supply, investigated the market potential of South Atlantic hake, handled and processed in such a way as to yield a product of high quality. One of the fishmongers invited to participate in the investigation expostulated: 'But there's plenty of cod this week!' C P Snow, referring to World War II, remarked that the main function of the scientist in government was to exercise foresight[47]. In its technical complexity, diversity and deep mysteries, the administration of the fisheries is not all that different from the administration of defence.

Technology, fisheries management and fisheries legislation

Earlier, we touched upon the topics of standards of construction of vessels, manning scales, qualifications of officers, hours and conditions of work, and the like. It is a little surprising that these have so rarely been invoked as criteria on which (and *inter alia*) coastal states would allow or forbid fishing vessels flying foreign flags to operate in waters within the jurisdiction of the coastal state. They are not often included in fisheries conventions dealing with bodies of water under the joint management of two or more states, nor are they often applied to the vessels of one state landing their catches of fish in the ports of another. It is surprising, because rules and regulations in such matters have a significant effect on the costs of fishing, and hence on the terms of competition between national fleets, and thus on attitudes as to what is overfishing (as the fisherman sees it) and what is not.

Often, in the drafting of fisheries law for developing countries, these topics are omitted completely, nor is mention always made of how responsibilities are to be divided and shared between the fisheries department and the authorities responsible for maritime law, ports and harbours, food standards, health and so on. Indeed, as implied earlier, such matters can give rise to problems in developed as well as developing countries, and are not always resolved in the best way from the point of view of the fishery industry or the consumer of fish. Partly for the historical reasons outlined earlier, departments of fisheries are not always well organized to keep themselves informed of developments in all scientific and technical fields relating to the wellbeing and prosperity of the industry, nor to avail themselves of high-level, well-informed advice on other dpartments' proposals, decisions and actions. Sometimes the expertise already exists within the fisheries service, sometimes it does not; and it may be as well to repeat here that there is no such person as a 'fisheries expert' in the sense of a single individual who can comprehend in detail all these various scientific and technical matters. Once again, the need is to coordinate and make best use of the talents of scientists, engineers and other technologists, economists, lawyers and others, in consultation with active fishermen, fish processors and distributors, if the full benefits are to be derived from the fisheries. Other aspects of the interaction of technology and fisheries management, including possibilities for new initiatives, have been noted in Chapter 4.

6
Technical development and world food supplies

The potential importance of the fisheries

Potential world production of species of more or less familiar size and form is believed to be about half as much again as present landings, but if we include the so-called 'unconventional' resources – the Antarctic krill, the mesopelagic species and the large oceanic squids – the potential is several times the present landings[6].

In many countries the fishery industry is at present of only minor importance to the national economy, although it may be of considerable importance locally. In some cases the size of the national industry will be too small to justify maintaining a permanent capability, in the public sector, in all the various areas of science, technology, law, marketing, economics and other specialisms that are relevant to the fisheries.

At the same time, because of the unfamiliarity of the fisheries industry to most people, and its economic, technical and environmental peculiarities, it tends to attract less attention than its potential would seem to justify, when plans are being formulated for economic development and investment. A high official of one of the development banks once remarked that, to a banker, the fisheries are strange and difficult country. This is not surprising, when it is remembered how little control we have over the weather, the changes in the ocean climate, the abundance and movement of the fish, and the activities of other fishermen. A farmer has collateral in the form of land, herds and crops in the ground; a fishing craft is a wasting asset: the real guarantee of the repayment of any loan is the skill of the fisherman. Often, the response of the officials in government and international agencies to a proposal for a new initiative in the fisheries is to ask: 'Will it pay?' The proposer may be forgiven for pointing out that, if he were sure it would pay, he would not be talking to a development agency but to his own bank manager. Not only does the all-too-apparent high risk of fishing enterprises ensure that the fisheries receive less recognition as valuable sources of food than their potential would seem to justify, but also the unfamiliarity of the technology to most decision-makers reinforces this attitude.

Technical assistance

There may therefore be a stronger than usual case for a well-organized and well-equipped system of international assistance in the fisheries, at both the policy-making and the technological levels. This cannot function effectively, however, unless the countries receiving the assistance possess a permanent cadre of professional people who are not only capable of identifying the areas of expertise in which assistance is required and of monitoring its execution, but who can also organize the necessary local participation in practical work in the field, especially at sea, and including extension work. Fisheries development is not merely or mainly a matter of procuring better vessels, equipment and processing plants, as all too often seems to be assumed: it must be fostered by a cadre of experienced fishermen, managers and technologists.

Many of the elements of a system of international technical assistance that could assist greatly in achieving this state of affairs in the developing countries already exist and a great deal of useful work has already been done. Much of the effort, however, has been expended upon resource surveys; most of the remainder has been more-or-less straightforward transfer of technology. This may no longer be adequate to meet the situation that now exists. As the production of fish increases, the stocks nearest to the densely-populated areas of the world tend to be the first to become fully exploited, always provided that they are composed of species acceptable to the local consumers, or are of species in heavy demand in the more affluent countries – as for example tunas, crustaceans and species which can be reduced to fish meal. There may come a time, however, when supplies of food fish may have to be sought in sea areas more remote from the main centres of population. There is also a requirement that systems be developed for harvesting and utilization of species that can be caught cheaply, in bulk, for feeding low-income groups in inland areas:

Plate 19 Increased fish supplies for local consumption in a developing country:

Plate 19(a) Awaiting the return of a kattumaram on a beach in Tamil Nadu (*Photo: BOBP*)

Plate 19(b) Shrimp trawlers have been re-deployed on to fishing for local consumption in Tamil Nadu (*Photo: BOBP*)

Plate 19(c) The increased supplies from the trawlers may require development of better means of marketing and distribution *(Photo: BOBP)*

the various stocks of small, oily pelagic shoaling species, some of which are at present exploited for reduction to meal and oil, seem to be the first choice for this purpose. Supplies of fish meal might be maintained by exploiting even more remote stocks, or species less attractive for direct human consumption, such as for example the mesopelagic species of the northern Arabian Sea. All of these possibilities present technical problems, the solutions to which are not to be found ready-made in other parts of the world: they require deliberate, directed programmes of technical development, supported by scientific investigation, and including pilot-scale projects to supply products for test-marketing.

These are only the most prominent examples of a general phenomenon: more and more the solution of problems in fisheries development, or the realization of new opportunities, requires local effort by way of technical development, to adapt methods and equipment, already in use elsewhere, to local circumstances. In some cases, they require genuine innovation. The development of improved beach-landing craft for the east coast of India, discussed earlier, is one example.

The place of high technology

The harvesting of such resources as the large oceanic squids, some of the mesopelagic fishes and the krill of the Southern Ocean requires high technology, special skills, techniques of fleet management and a degree of organization which are outside the experience of any nation except those who have already engaged in distant-water fishing in high latitudes. These nations, however, are not in urgent need of new sources of supply of animal protein, except perhaps in the form of fish meal, whereas some of the developing nations might be more interested in additional supplies for their human populations or livestock. Thus we may have a situation where some nations possess the capability of harvesting these resources, but other, poorer nations may wish to consume the products. Perhaps the new international economic order will result in arrangements satisfactory to all concerned.

Development options for food production

Whether, and to what extent, it is desirable to foster the development of the fisheries so as to provide additional supplies of food, or to relieve the intensity at which the land must be cropped, are questions that have not yet been the subject of much deliberate study, except in terms of efficiency of utilization of energy from fossil fuels, as compared with production of other kinds of animal protein. Such matters do not seem to be yet under regular review at the international level. Indeed such an approach is still uncommon at the national level, although Wikstrom has drawn attention, in one particular case, to the existence of a choice between two different options for future investment in fish production: further investment in a marine capture fishery that might then operate uncomfortably near the maximum sustainable yields; or investment in the cultivation of fish in fresh waters, lakes and reservoirs[48]. One of the factors to be taken into consideration is the present preference of many consumers in that particular country for marine fish; experience elsewhere suggests, however, that such preferences can change or be changed. Product development and marketing are areas where additional effort might be very worth while in many parts of the world.

For the last three or four years the possibilities of exploiting more fully the potential of the fisheries, pointed out by FAO[6], have received comparatively little attention, partly because of the consequences of rises in the real cost of energy and partly because of the preoccupation with 200-mile limits. Indeed, the adoption of extended fisheries limits and the subsequent exclusion of foreign distant-water fleets has led in some cases to a fall in the quantity and more especially in the quality of the landings (as was predicted by one or two technologists); this happened in at least one developed country as well as in some developing ones. The need for exchange of knowledge and experience in the field of handling and processing is just as great as it is in the fields of fisheries management and the technology of capture, but receives less attention.

The technical feasibility of harvesting those underexploited resources, the existence of which is already known, is less in doubt than the precise ways in which they should be utilized and the forms of the final products that will be most acceptable to the potential consumers. Where the underexploited resources are already known to be large, the most urgent need is not a more precise estimate of the biomass and ultimate potential, but the identification by practical trial of effective means of harvesting and utilization, on a sufficient scale to allow an assessment to be made of market acceptability and an estimate of costs of production and distribution. Only then can the potential be assessed, and such assessments would still need to be reviewed and revised from time to time in the light of changing circumstances and the advent of new technology.

Meanwhile, in the advanced fishing nations, there will be a continuing need for fisheries development organizations capable of providing an adequate response to changes in external circumstances or of assisting government and industry to attain the objectives of national fisheries policy. For the reasons noted earlier, these could at the same time provide the experienced professional nucleus of a more comprehensive system than at present exists for providing technical assistance and advice to the less developed fishing nations.

Qualified manpower

The number of engineers and other technologists actively engaged in fisheries development is now probably less than it was during the period of expansion from 1950 to 1970, when the projects were probably less technically difficult than some that might now be tackled (although the engineers were then by no means sufficiently experienced, or numerous enough, to help the industry avoid many false starts and unwise investments). There has not been sufficient recruitment to replace losses and, especially, to maintain the seagoing capabilities of the ageing development teams. Some of the organizations that developed the techniques of measurement, observation, and processing of information, and the arts of procurement, testing and demonstration of prototype methods and equipment, and which produced some of the more spectacular advances, are now run down or in reduced circumstances. The number of engineers engaged full time in fisheries research and development has always been small, as might be expected, in so far as their functions are only to suggest technical initiatives, and to act as technical intermediaries between user, supplier and government. In 1950 there may have been no more than a couple of dozen in the world as a whole; in the United Kingdom, which had one of the first and most active units engaged in technical development of fishing vessels and their equipment to the stage when the results were immediately applicable by industry, the number of professional engineers and operations analysts so engaged, and in the associated fisheries research establishments, was never much more than thirty, half of whom comprised the seagoing teams.

The constraint on the development of world fisheries during the next few years may not be a scarcity of fish, nor a shortage of men able and willing to go to sea in commercial fishing vessels, but a lack of engineers and technologists to develop appropriate systems of harvesting the resources, and also to develop acceptable forms of product whereby they can be effectively utilized.

7
Conclusion

To ensure adequate response to change, the fisheries industry needs a capability for adapting its methods, equipment and systems and for developing new methods, equipment and systems as may be necessary or desirable. Such a capability is also necessary if industry and government are to attain the objectives of national fisheries policy and ensure maximum benefits to the consumer, to the fisheries industry and to the nation as a whole, at least cost. It is desirable to investigate new fishing opportunities in a practical way, so as to allow appraisal of potential future costs and benefits.

There are few enterprises in the fishery industry itself capable of organizing and financing such development activities and those which are, have somewhat specialized or narrow interests and affiliations. Much has to be done by the public sector.

It is unprofitable to separate fisheries technology from fisheries management, which needs to comprise all public sector decisions relating to the harvesting and utilization of fishery resources. Fisheries management is the responsibility of governments and inter-governmental agencies and these must also, for a variety of reasons, take responsibility for much of the effort in fisheries development.

Much remains to be done to establish and foster adequate and well-integrated organizations that can respond to both national and international requirements and needs in the field of fisheries development and also to allow them to make their proper contribution to devising and establishing effective systems of fisheries management.

Catching, handling and processing, distribution and marketing, and conservation, all interact; action in one field can either mitigate or exacerbate the problems in one or more of the others, or facilitate a new approach. Those engaged in any particular field need not accept unquestioningly the current practice in the others, nor should they seek to impose solutions that are distinguished only by requiring no change in their own. The problems of establishing effective fishery management regimes are great; so, however, may be the variety of options that might be revealed by an appreciation based on a thorough, well-informed review of the options and opportunities in all the various fields of harvesting, utilization, marketing and conservation. The real problem may be the difficulty of discerning which combination of options will best serve national policy objectives.

Nevertheless we cannot hope to exercise much real control over the marine environment, and to that extent at least, a continued ability to respond to change, through technical versatility coupled with flexibility in the market, will be necessary for the future prosperity of the industry and the continuing satisfaction of the consumer.

References

1 Burgess, G.H.O.: The Curious World of Frank Buckland. John Baker, London, 1967.

2 For large fishing vessels, *see Lloyd's List,* published periodically by Lloyd's Register of Shipping, London. *See also* various national fishery statistics.

3 Anon. FAO Yearbooks of Fisheries Statistics. FAO, Rome.

4 Eddie, G.C.: Technical Development in the Fishing Industry – the Work of the White Fish Authority. *Royal Institution of Naval Architects, Spring Meeting 1971* Paper 7. 13p. 13 refs.

5 Clark, G.: World Prehistory, an Outline. Cambridge University Press, 1961.

6 Anon. The Potential of the Fisheries to Provide Increasd Food Supplies for the Developing Countries and the Requirements for Investment. *FAO Fisheries Circular* No. 343, Rev. 1. Rome, 1977. 21pp.

7 Straus, L.G., Clark, G.A., Altuna, J. and Ortea, J.A.: Ice-Age Subsistence in Northern Spain. *Scientific American* Vol. 242 No. 6 June 1980 pp 120–129.

8 Culican, W.: The First Merchant Venturers, *ed.* Stuart Piggott. Thames and Hudson, London, 1966.

9 Modiano, M.: Corinthian Cannery. *Sunday Times,* London, No 8219, 10 January 1982, p. 15.

10 Piggot, S.: Ancient Europe. Edinburgh University Press, 1965.

11 Wilson,–.*ed.*: Medieval Merchant Venturers. Methuen, 1954.

12 Anon. *Scottish Fisheries Bulletin* No. 20, 1963, p. 26. Department of Agriculture and Fisheries for Scotland, Edinburgh.

13 Mattingly, G.: The Defeat of the Spanish Armada. Jonathan Cape, London, 1959.

14 Eddie, G.C.: The Expansion of the Fisheries: New Applications of Mechanical Power in Man's Oldest Industry. James Clayton Lecture. *Institution of Mechanical Engineers,* London. *Proceedings* 1970-71 Vol. 185 42/71.

15 Kurien, J. and Willmann, R.: Economics of Artisanal and Mechanized Fisheries in Kerala. FAO/UNDP Project for Promotion of Small-Scale Fisheries in South Asia RAS/77/044, Working Paper No. 34. Madras, July 1982. 112pp.

16 Bergstrom, M.: A Pilot Survey of Fish Landings at Three Sites in Bangladesh. FAO/SIDA Project for the Development of Small-Scale Fisheries in the Bay of Bengal (BOBP), Madras, India. In preparation (1982).

17 Eddie, G.C. and Nathan, M.T.: Pre-Feasibility Study of a Floating Fish Receiving and Distribution Unit for Dubla Char, Bangladesh. Working Paper published by BOBP (see ref. 16), Madras, 1980.

18 Marr, G. Alan: Personal communication, 1968. J. Marr and Son Ltd, Hull, England.

19 Gulbrandsen, O., Gowing, G.P. and Ravikumar, R.: Technical Trials of Beachcraft Prototypes in India. Working Paper published by BOBP (see ref. 16), Madras, 1980.

20 ———: Further Trials of Beach Landing Craft. Working Paper to be published by BOBP (see ref. 16) 1982.

21 Gillet, P.: Small-Scale Fishery Development Projects at Muttom, Kanyakumari District – a Case Study. *In* Proceedings of Seminar on the Role of Small-Scale Fisheries and Coastal Acquaculture in Integrated Rural Development, Madras, December 6–7, 1978. *Central Marine Fisheries Research Institute of India* 1/1978/CMFRI. Cochin, 1978.

22 Carleton, C.: The Planning of Development Projects in Small-Scale Fisheries. *In* Proceedings of the FAO Expert Consultation on Fish Technology in Africa, Casablanca, Morocco, 7-11 June 1982, pp 275–279. FAO FIIU/R268 Suppl., Rome, 1982.

23 Hunter, A. and Eddie, G.C.: Fishing Vessel Development. *Institute of Marine Engineers, Transactions* 71(6) June 1959.

24 Chaplin, P.D. and Haywood, K.H.: Operational Research Applied to Stern Freezer Trawler Design. *Institute of Marine Engineers,* London. Specialist meeting at Grimsby, March 1968.

25 Schaerfe, J.: The German One-Boat Midwater Trawl. *Fishing News International.* Published in parts, July 1969 – December 1969.

26 Blum, P.: Bigger Trawlers, Bigger Yields: a Result of Increased GRT or HP? *Die Fischwirtschaft* Vol. 7 August 1955, p. 290.

27 Troadec, J.-P.: Introduction a l'Aménagement des Pêcheries: Interêt, Difficultés et Principales Méthodes. *FAO Fisheries Technical Paper* 224, 65pp., Rome, 1982. (English and Spanish versions in preparation).

28 Cooling, W.A.: Personal communication, 1955. Northern Trawlers Ltd., Grimsby, England.

29 Steinberg, M.A.: Past, Present and Future Methods of Utilization *in* Advances in Fish Science and Technology, ed. J.J. Connell *et al.* pp 34–48. Fishing News Books, Farnham, Surrey, 1980.

30 Ruckes, E.: Fish Marketing Management. Paper delivered to Workshop on Management of National Fishery Development Programmes. White Fish Authority, Hull, England 1981. (Now Sea Fish Industry Authority, Industrial Development Unit, Hull, England).

31 Rothschild, B.J.: First Annual David H. Wallace Memorial Lecture on Marine Living Resources. Marine Technology Society Journal Vol. 14 No. 5 pp 5–11 1980.

32 ———: More Food From the Sea? *Bioscience* Vol. 31 No. 3 March 1981 pp 216-222.

33 Pickering, W.: Sources of Finance. Paper to Workshop on Management of National Fishery Development Programmes, WFA, Hull, England, 1981 (see ref. 30).

34 Pajot, G. and Das, T.K.: Trials in Bangladesh of Large-Mesh Drift Nets of Light Construction. BOBP/WP/12 Madras 1981. (see ref. 16).

35 (FAO) Report on the Expert Consultation on the Regulation of Fishery Effort, FAO, Rome, January 1983. In preparation.

36 Anon. Report of the ACMRR Working Party on the Scientific Basis of Determining Management Measures, Hong Kong, December 1979. *FAO Fisheries Report* No. 236, Rome, 1980.

37 Robertson, A.: Personal communication, 1964. Department of Engineering. University of Bristol.

38 Bennet, R., Cartwright, G. and Siddle, W.: Measurement of Loads and Powers on the Trawl Winch of D.E.T. JUNELLA. *White Fish Authority Technical Memorandum* No. 16 (Restricted Circulation). (Now Sea Fish Industry Authority, Industrial Development Unit, Hull, England).

39 Anon. *See* Reports of the Food Investigation Board, Department of Scientific and Industrial Research (annual) and *Torry Research Station Memoirs* (occasional), HMSO, London, 1929 *et seq.*

40 Lochridge, W.: Mechanization in Fishing Vessels. *Transactions of the Institution of Engineers and Shipbuilders in Scotland* Vol. 99, p 511. Glasgow, 1955–56.

41 Anon. Progress Report on the Trawl Gear Development Project. White Fish Authority, July 1962. (Now Sea Fish Industry Authority, Industrial Development Unit, Hull, England).

42 Iles, D.: Personal communication, 1970. Department of Fisheries, Environment Canada.

43 Crutchfield, J.A.: Economic and Political Objectives in Fishery Manage-

ment *in* World Fisheries Policy, ed. B.J. Rothschild, University of Washington Press, Seattle and London, 1972, pp 74–89.

44 Smith, V.L.: The Primitive Hunter Culture, Pleistocene Extinction, and the Rise of Agriculture. *Journal of Political Economy* Vol. 8 No. 41, 1975.

45 *See* for example, *Proceedings of the Royal Institution of Naval Architects* in the years 1980–82, and corespondence in the Institution's Journal *(The Naval Architect)* during the same period.

46 Eddie, G.C.: Institutions and Organisations in Fisheries Development. Paper to Workshop on the Management of National Fisheries Development Programmes, Hull, England, 1981. (Sea Fish Industry Authority, Industrial Development Unit, Hull, England).

47 Snow, C.P.' Science and Government: the Godkin Lectures at Harvard University, 1960. Oxford University Press 1961.

48 Wikstrom, U.: Personal communication, 1981. Acquaculture Programme, Department of Fisheries, FAO, Rome.

Other books published by
Fishing New Books Ltd

Free catalogue available on request